KAPLAN

Test Prep and Admissions

MATH SOURCE

The Smarter Way to Learn Math

by

Catherine V. Jeremko

Colleen M. Schultz

Simon & Schuster

NEW YORK · LONDON · SYDNEY · TORONTO

Kaplan Publishing
Published by SIMON & SCHUSTER
1230 Avenue of the Americas
New York, NY 10020

Editorial Director: Jennifer Farthing
Project Editor: Sheryl Gordon
Production Manager: Michael Shevlin
Interior Design and Page Layout: Dave Chipps
Cover Design: Cheung Tai & Mark Weaver

Manufactured in the United States of America.
Published simultaneously in Canada.

August 2005
10 9 8 7 6 5 4 3 2

ISBN-13: 978-0-7432-5159-4
ISBN-10:   0-7432-5159-8

Contents

About the Authors

Catherine V. Jeremko is a certified secondary mathematics teacher in New York State. She is the author of mathematics test prep and review materials, as well as a manuscript reviewer for Kaplan Publishing. She currently teaches seventh grade mathematics at Vestal Middle School in Vestal, New York. Ms. Jeremko is also a teacher trainer for both mathematical pedagogy and the use of technology in the classroom. She resides in Apalachin, New York, with her three daughters.

Colleen M. Schultz is certified in both secondary mathematics and elementary education in New York State. She is the author of mathematics test prep and review materials, as well as a manuscript reviewer for Kaplan Publishing. She currently teaches eighth grade mathematics at Vestal Middle School in Vestal, New York, where she recently served as a teacher mentor for the Vestal School District. Mrs. Schultz is also a teacher trainer for both mathematical pedagogy and the use of technology in the classroom. She resides in Binghamton, New York, with her husband and two children.

Introduction

Dear Reader,

Do positive and negative numbers make your head swim? Do you use ten steps to solve a problem when you could have used three? Or are you just confused about the difference between a trapezoid and a rectangle? In any case, you picked up this book, which means you want to improve your math skills—a smart decision. Why? **Because being able to apply math skills and concepts will get you places**—on standardized tests, in the workplace, and in everyday life. Strong math skills can help you get ahead, get promoted, or simply save you time when solving the problems you face daily.

The problem is, improving your skills in mathematics is often a boring, tedious process. But it doesn't have to be. Allow us to introduce the **Building Block Method.**

This teacher-approved method was devised by the experts at Kaplan to make learning mathematics as painless as possible. Like any other book, you'll begin with the basic math principles and concepts you need to master. But then you'll move on to a systematic routine for memorization that uses your own real-life situations as practice exercises—a convenient way to learn math while going about your everyday life.

There's no smarter way to learn. So get started—good math skills are only a few building blocks away!

HOW TO USE THIS BOOK

Whether you read *Math Source* from start to finish or only want to brush up on certain topics, this book will provide you with a systematic method for improving your skills. Each chapter contains five key components.

1. Building Block Quiz

You'll begin each chapter with a short quiz. The first three questions will cover material from earlier lessons, so if you get these questions wrong, it's time to go back and review! The next seven questions will test your knowledge of the material to be covered in the upcoming chapter. Taking this quiz helps reinforce what you've already learned while targeting the information you need to focus on in each chapter. Plus, you'll get even more review from the answer explanations, which tell you exactly why each answer choice is right or wrong.

2. Detailed Lesson

Each chapter focuses on one specific math topic. We'll explain the concepts involved in detail, provide lots of relevant examples, and offer strategies to help you remember what you need to know. In the process, we help you review the fundamentals from previous chapters.

3. Plentiful Practice

Here's a phrase you'll hear over and over throughout this book: *Repetition is the key to mastery*. So be prepared to practice, practice, practice! You'll find everything from simple matching exercises to exercises that ask you to apply the skills you're learning to practical, real-life situations. By "learning from all sides," so to speak, you're much more likely to retain the information. And remember, don't start a new chapter if you haven't mastered the earlier material – you'll be building on a weak foundation.

4. Summary

Each chapter concludes with a bulleted summary for quick review of important key facts you should take from the lesson.

5. Chapter Test

At the end of each chapter, you'll take a post-test to help you practice what you've learned and assess how well you've learned it. The tests cover material in that chapter, plus key concepts from previous chapters. The chapter test will help you make sure you have mastered the material in that chapter before you move on to the next lesson.

A strong building needs a strong foundation. To improve your skills in math, you need to begin with a strong foundation, too—a solid understanding of basic math concepts. The 12 chapters in this book are arranged from basic skills to more advanced topics, beginning with the properties of numbers, so everything else builds upon a firm base.

With a system as easy as this, good math skills are well within your reach. All you have to do is take the first step. Good luck!

CHAPTER 1

Properties of Numbers

BUILDING BLOCK QUIZ

Start off your math review with a building block quiz of 10 questions. This quiz will be a warmup to chapter 1, which covers the concepts of number sets, order of operations, and number properties.

1. Which set below features the natural numbers?

   (A) $\{\ldots-5, -4, -3, -2, -1, 0, 1, 2, 3, 4, 5 \ldots\}$

   (B) $\{0, 1, 2, 3, 4, 5 \ldots\}$

   (C) $\{1, 2, 3, 4, 5 \ldots\}$

   (D) $\{0, \dfrac{1}{4}, \dfrac{1}{2}, \dfrac{3}{4}, 1, \dfrac{5}{4}, \dfrac{3}{2} \ldots\}$

   (E) $\{1, \dfrac{3}{2}, 2, \dfrac{5}{2}, 3, \dfrac{7}{2}, 5\}$

2. Which number below is not in the set of rational numbers?

   (A) $\dfrac{5}{4}$ (B) $\dfrac{2}{3}$ (C) π (D) $7.\overline{8}$ (E) 4.125

3. Simplify: $10 - 8 + 3$

   (A) 21   (B) -1 (C) 15   (D) 1   (E) 5

4. Simplify: $20 + 6^2 \div 4$

   (A) 14   (B) 29   (C) 169   (D) 8   (E) 2

5. Simplify: $\dfrac{21 + 7 \times 2}{7}$

    (A) 5   (B) 35   (C) 19   (D) 8   (E) 23

6. Simplify: $30 - 12 \div 3 + 9$

    (A) 17   (B) 15   (C) 29   (D) 35   (E) 1.5

7. Which example below demonstrates the associative property?

    (A) $7 + 3 + 29 = 7 + (3 \times 29)$

    (B) $7(29) = (7 + 20) \times (7 + 9)$

    (C) $7(29) = 7(20) + 7(9)$

    (D) $7 + 29 + 3 = 3 + 7 + 29$

    (E) $7 + (3 + 29) = (7 + 3) + 29$

8. The statement $6.4 + 5.4 + 3.6 = 6.4 + 3.6 + 5.4$ is an example of the

    (A) associative property of addition

    (B) commutative property of addition

    (C) distributive property of multiplication over addition

    (D) order of operations

    (E) none of the above

9. Which example below demonstrates the distributive property of multiplication over addition?

    (A) $12 + 19 + 8 = 12 + 8 + 19$

    (B) $19 + 12 + 8 = 19 + (12 + 8)$

    (C) $8 \times 29 = (8 + 20) \times (8 + 9)$

    (D) $8 \times 29 = (8 \times 20) + (8 \times 9)$

    (E) $8 \times 29 = (29 \times 8) + (29 \times 1)$

10. Which of the following statements is NOT true?

   (A) 0 is an even number

   (B) 0 is a positive number

   (C) 0 is a whole number

   (D) 7 is a rational number

   (E) $3.\overline{8}$ is a rational number

Answers and Explanations

1. C

This problem, along with problem 2, tests your knowledge of the classification of numbers. Only choice (C) features the natural numbers, which are the counting numbers excluding 0: 1, 2, 3, 4, etc. Choice (A) features the integers. Choice (B) is the set of whole numbers. Choice (D) includes some rational numbers, and choice (E) is a set of rational numbers.

2. C

Pi (π) is a constant number that is irrational. An irrational number cannot be exactly expressed as a fraction because in decimal form it neither terminates nor repeats. Rational numbers are

numbers that can be written in the form $\frac{a}{b}$, where a and b are integers and $b \neq 0$.

Therefore, choices (A) and (B) are by definition rational numbers.

In choice (D), $7.\overline{8} = 7\frac{8}{9} = \frac{71}{9}$, a rational number. In choice (E),

$\frac{4,125}{1000} = 4\frac{1}{8}$, also a rational number.

3. E

Questions 3, 4, 5, and 6 test your knowledge of the order of operations. First evaluate $10 - 8$, since subtraction is the first operation when reading from left to right. Then add: $2 + 3 = 5$.

4.   B

Remember the order of operations and evaluate the exponent first:
$6^2 = 36$. Next, evaluate division: $36 \div 4 = 9$. Finally, evaluate addition:
$20 + 9 = 29$.

5.   A

A fraction bar is a grouping symbol; first evaluate the numerator (the top
of the fraction) and then the denominator (the bottom of the fraction).
For the numerator, perform multiplication first to get $7 \times 2 = 14$. Next
comes addition: $14 + 21 = 35$. The final operation is dividing the
numerator by the denominator: $35 \div 7 = 5$.

6.   D

According to the order of operations, division is performed first:
$12 \div 3 = 4$. Next, moving from left to right, subtract this 4 from 30:
$30 - 4 = 26$. Finally, $26 + 9 = 35$.

7.   E

Questions 7, 8, and 9 concern the properties of numbers. The
associative property of addition says that if the grouping of addends
changes, the sum remains the same. This is shown in choice (E),
where there is only addition, and the grouping is the only operation
changed. Choices (A) and (B) have both addition and multiplication as
operations, and are not even true statements. If you simplify the
equation in choice (B) by using the order of operations, you get $203 =
27 \times 16$, or $203 = 432$, which is false. Choice (C) is an example of the
distributive property of multiplication over addition. Choice (D) is an
example of the commutative property of addition, in which the order of
the addends is changed.

8.   B

The example shown demonstrates the commutative property, which
states that changing the order of addends does not affect the sum.
Choice (A), the associative property, states that changing the grouping
of addends does not affect the sum. Choice (C), the distributive
property, indicates that multiplication distributes over addition. Choice
(D), the order of operations, are the rules of precedence for
performing arithmetic operations.

9.   D

Choice (D) shows how multiplication distributes over addition. Choice (A) is an example of the commutative property of addition, in which the order of the addends is changed. Choice (B) is an example of the associative property of addition, in which the grouping of the addends is changed. Choice (C) is false because it indicates that addition would distribute over multiplication, which is not true. Choice (E) is not an example of any property of numbers; it is not a true statement. If you simplify the equation in choice (E) by using the order of operations, you get $232 = 232 + 29$, or $232 = 261$, which is false.

10.   B

This question deals with the sets of numbers. Be careful! This problem asks which statement is NOT true. Choice (A) is true; 0 is an even number since it is divisible by 2. Choice (B) is the only choice that is NOT true. The number 0 is neither positive nor negative. Choice (C) is true—0 is in the set of whole numbers by definition. Choice (D) is true. A rational number is a number that can be expressed as a fraction, and $7 = \frac{7}{1}$.

Choice (E) is also correct; any decimal that either terminates or repeats is a rational number.

WORKING WITH NUMBERS

In this first chapter you will review the basic facts of working with numbers, including the different classifications of numbers, the order of operations when simplifying numerical expressions, and the three basic properties of numbers. This introductory material will form the foundation of *Math Source*.

Sets of Numbers

The **real numbers**, all of the numbers that you encounter each day, are classified into various sets and subsets. All real numbers are either **rational numbers** or **irrational numbers**.

> **MATH SPEAK**
>
> The **rational numbers** are the numbers that can be expressed as a fraction in the form $\frac{a}{b}$, where a and b are integers and $b \neq 0$.

A rational number can take various forms: 5 can also be written as $\frac{5}{1}$; 1.2 can be written as $\frac{12}{10}$ or $\frac{6}{5}$; −7 can be written as $\frac{-7}{1}$; and $0.\overline{3}$ can be written as $\frac{1}{3}$. When written as a decimal, a rational number is either *terminating* or *repeating*. Fractions and decimals will be covered in depth in chapter 3.

> **MATH SPEAK**
>
> The **rational numbers** can be classified into subsets:
>
> The **natural numbers**, also known as the counting numbers: { 1, 2, 3, 4, 5, 6, 7... }.
>
> The **whole numbers** are the natural numbers, plus 0: { 0, 1, 2, 3, 4, 5, 6, 7... }.
>
> The **integers** are the whole numbers and their opposites: { ...−3, −2, −1, 0, 1, 2, 3... }.

Some things you should know about integers:

- All numbers greater than 0 are positive numbers.
- All numbers less than 0 are negative numbers.
- Zero is neither positive nor negative.
- The even integers are divisible by 2, and include the number 0.
- The odd integers are not divisible by 2.

The integers will be covered in more depth in the next chapter.

The irrational numbers cannot be expressed as fractions. Some examples are $\sqrt{2}$, $\sqrt{3}$, $\sqrt{5}$, and π. When irrational numbers are expressed as decimals, they are *non-terminating* and *non-repeating*. Some examples of irrational decimals are: 3.141592653589793238462643383279502884197, or 7.030030003…, or 8.81811811181111…

Try the matching exercise below to test your understanding of the number sets.

Practice 1

For each set of numbers in the first column, find the letter that corresponds with correct set classification name. Choose the most descriptive name for the set. Answers and explanations are located at the end of the chapter.

1. $\{\sqrt{7}, \sqrt{11}, \pi, 5.1101001000...\}$ A. rational numbers

2. $\{ ...-3, -1, 1, 3... \}$ B. whole numbers

3. $\{ 0, 1, 2, 3, 4... \}$ C. natural numbers

4. $\{ -3, .4, -1.5, 0, 6.3 \}$ D. irrational numbers

5. $\{ 1, 2, 3 ... \}$ E. odd integers

The Order of Operations

You encounter formulas everyday in life. When traveling abroad, you may need to convert a temperature from degrees Celcius to degrees Fahrenheit to better understand it. You many need to calculate the perimeter or area of your backyard to build a fence. Formulas exist for all kinds of situations.

When simplifying a mathematical expression after you have plugged values into your formula, you do not simply work from left to right, as you do when you read a book. Just as there are rules for driving an automobile, there are rules for order when performing arithmetic operations. There is a predetermined **order of operations** used to evaluate expressions. Perhaps you remember the mnemonic for remembering the order of operations: PEMDAS. Some of you may have the used memory tool "Please Excuse My Dear Aunt Sally" to recall the correct order.

> # REMEMBER THIS!
>
> The order of operations is:
>
> P   Parentheses (grouping symbols)
>
> E   Exponents
>
> MD Multiply and divide from left to right
>
> AS Add and subtract from left to right

Active participation is an important part of learning, and repetition is the key to mastery! Try these examples to test your knowledge of the order of operations.

Practice 2

For questions 6–9, answer true or false for each statement. Answers and explanations are located at the end of the chapter.

6. **T F**   In a numeric expression, addition is always performed before subtraction.

7 **T F**   Parentheses are evaluated before exponents.

8. **T F**   In the expression $57 - 32 + 8$, subtraction is the last operation performed.

9. **T F**   In the expression $700 \div 14 \times 2$, you would first multiply 14 by 2.

10. Simplify: $4^2 - (20 \div 4 \times 2)$

11. Simplify: $(15 - 12)^3 \div 9 \times 3$

12. Simplify: $94 - 2(4)^2 + 14$

13. Simplify: $1,200 \div 10^2 - 5 \times 2$

The "P" in PEMDAS stands for parentheses, or grouping symbols. Grouping symbols include parentheses, brackets, the absolute value symbol, and a fraction bar. So to simplify $\dfrac{18 + 10^2 - 4 \times 2}{20 - 27 \div 3}$, treat the fraction bar as a grouping symbol and first evaluate the top (the numerator) then the bottom (the denominator). Then you will divide for the final step.

To simplify the numerator, first simplify your exponent: $10^2 = 100$. Second, multiply 4 times 2 to get 8. The top is now $18 + 100 - 8$. Evaluate from left to right: $118 - 8 = 110$. To simplify the denominator, first divide 27 by 3 to get 9. Then subtract: $20 - 9 = 11$. Finally, divide 110 by 11 to get 10.

In the order of operations, a radical sign is evaluated on the same level of priority as an exponent. To simplify $550 - \sqrt{9 \times 4} \times 3$, you would evaluate the radical first. Under the radical sign, multiply 9 times 4 to get 36. The square root of 36 is 6. Now the problem reads $550 - 6 \times 3$. Multiply 6 by 3 next to get $550 - 18$, for a final value of 532.

When plugging numbers into formulas, a working knowledge of the order of operations is essential. For example, to convert a temperature from degrees Fahrenheit to degrees Celsius you use the formula $C = \dfrac{5}{9}(F - 32)$, where F is the degrees in Fahrenheit and C is the degrees in Celsius. If you have a temperature of 77 degrees Fahrenheit, and you want to know the equivalent degrees in Celsius, substitute in 77 for F in the formula: $C = \dfrac{5}{9}(77 - 32)$.

First, subtract 32 from 77, because parentheses are evaluated first: $C = \dfrac{5}{9}(45)$. Now, multiply $\dfrac{5}{9}$ by 45 to get 25 degrees Celsius.

NUMBER PROPERTIES

There are common properties of numbers that are frequently used to make adding and multiplying easier. You most likely use these properties without even realizing it when you do mental arithmetic or when you add a column of numbers. These properties give you the "license" to change the order of operations in certain situations. In addition to making addition and multiplication of number terms easier to calculate, these three properties are frequently used in solving algebraic equations, as explained in chapter 12.

The Commutative Property of Addition

> **MATH SPEAK**
>
> The **commutative property of addition** states that changing the order of the addends in a sum does not change the sum.
>
> $a + b = b + a$, where a and b are any real numbers.

For example:

$$12.3 + 6.9 + 7.7 = 12.3 + 7.7 + 6.9$$

The order of operations would dictate that 12.3 would first be added to 6.9. But the addition is easier if you first add 12.3 to 7.7, because the sum will equal a whole number. The commutative property gives you this freedom.

The Commutative Property of Multiplication

> **MATH SPEAK**
>
> The **commutative property of multiplication** states that changing the order of the factors in a product does not change the product.
>
> $a \times b = b \times a$, where a and b are any real numbers.

For example:

$$2 \times 8 \times 5 \times 7 = 2 \times 5 \times 8 \times 7$$

If you scan a group of factors to find sub-products to equal 10, 100, or 1,000, it is easiest to multiply these factors first. The commutative property allows you to make these changes to the order of operations.

The Associative Property

The associative property also pertains to either the addends in a sum or the factors in a product.

> **MATH SPEAK**
>
> The **associative property of addition** or **multiplication** states that changing the grouping (parentheses or brackets) of addends in a sum or the grouping of factors in a product does not change the resulting sum or product.
>
> $a + (b + c) = (a + b) + c$, where a, b, and c are any real numbers.
> $a \times (b \times c) = (a \times b) \times c$, where a, b, and c are any real numbers.

For example, to add $9.8 + (10.2 + 6.1) + 4.9$, the order of operations would call for you to evaluate inside the parentheses first and then to add from left to right. But the sum of 9.8 and 10.2 is 20, and the sum of 6.1 and 4.9 is 11. The associative property allows you to change the grouping by adding: $(9.8 + 10.2) + (6.1 + 4.9)$, to get $20 + 11 = 31$. Notice in the example that the order of the addends did not change, just the grouping.

To multiply, consider the expression $7 \times 20 \times 5 \times 8$. Notice that $20 \times 5 = 100$, so change the grouping to make the multiplication easier: $7 \times (20 \times 5) \times 8$. Now, evaluate from left to right: $7 \times 100 = 700$. Finally, $700 \times 8 = 5,600$.

You can also use a combination of the properties. For example, to simplify the expression $2.1 + 8.07 + 7.9 + 24.93$, scan the addends and recognize that $(2.1 + 7.9)$ and $(8.07 + 24.93)$ will produce whole numbers. Use the commutative property to get $2.1 + 7.9 + 8.07 + 24.93$. Then use the associative property to get $(2.1 + 7.9) + (8.07 + 24.93)$. Now the addition is easy to finish: $10 + 33 = 43$.

Practice 3

Practice your knowledge of these properties. For each example below answer commutative, associative, or both. Answers are located at the end of the chapter.

14. $7 + 11 + 3 = 7 + 3 + 11$

15. $12 \times 25 \times 4 = 12 \times (25 \times 4)$

16. $16.4 + 7.9 + (0.1 + 3) = 16.4 + (7.9 + 0.1) + 3$

17. $17.3 + 15 + 12.7 + 30 = (17.3 + 12.7) + (15 + 30)$

The Distributive Property

The distributive property involves two operations: addition and multiplication or subtraction and multiplication.

> **MATH SPEAK**
>
> The **distributive property of multiplication over addition or subtraction** states that multiplication distributes over addition and subtraction.
>
> $a \times (b + c) = (a \times b) + (a \times c)$, where a, b, and c are real numbers
>
> $a \times (b - c) = (a \times b) - (a \times c)$, where a, b, and c are real numbers

For example, if you want to multiply 16 by 8, you may not know the multiples of 16. However, you do know the multiples of 10 and 6. The distributive property allows you to rewrite 8×16 as $8 \times (10 + 6)$, or simply $8(10 + 6)$. Because multiplication distributes over addition, this problem becomes $(8 \times 10) + (8 \times 6)$, which can be evaluated as $80 + 48 = 128$. Likewise, 8×16 could be written as $8 \times (20 - 4)$ or $(8 \times 20) - (8 \times 4) = 160 - 32 = 128$.

You can also use the distributive property in reverse. For example, if you were instructed to simplify $(12 \times 6.4) + (12 \times 3.6)$, order of operations would have you evaluate parentheses first, which involves decimal multiplication. If you notice that both terms are multiplied by 12, use the distributive property to "factor out" the 12: $(12 \times 6.4) + (12 \times 3.6) = 12 \times (6.4 + 3.6) = 12 \times 10 = 120$.

Practice 4

To practice using the distributive property, fill in the blanks below.

18. $9(50 + 6) = ($ _____ $\times 50) + (9 \times 6)$

19. $(18 \times 24) + (18 \times 76) =$ _____ $(24 + 76)$

20. $(15 \times 6.6) + (15 \times 3.4) = 15(6.6 +$ _____ $)$

21. $12(30 + 14) = ($ _____ $\times 30) + (12 \times 14)$

SUMMARY

In conclusion, here are five main points about number properties that you should take from this chapter:

- The correct order of operations is parentheses, then exponents, then multiplication and division, and then addition and subtraction (PEMDAS).
- For order of operations, multiplication and division are performed from left to right; addition and subtraction are also evaluated from left to right.
- The commutative property states that changing the order of addends or factors does not change the sum or product.
- The associative property states that changing the grouping of addends or factors does not change the sum or product.
- The distributive property states that multiplication distributes over addition or subtraction.

Practice Answers and Explanations

1.   D

The irrational numbers cannot be expressed as fractions.

2.   E

The integers are the whole numbers and their opposites. Odd numbers cannot be divided evenly by 2.

3.   B

The whole numbers are the natural numbers, plus 0.

4.   A

The rational numbers are the numbers that can be expressed as fractions.

5.   C

The natural numbers are also known as the counting numbers.

6.   False

Addition and subtraction are evaluated from left to right, as they are encountered in a problem.

7.   True

Parentheses are always evaluated before exponents.

8.   False

Addition will be evaluated last.

9.   False

Division will be evaluated first.

10.   6

Parentheses are evaluated first. Inside the parentheses, evaluate division and then multiplication: $20 \div 4 = 5$; $5 \times 2 = 10$. Next, evaluate the exponent: $4^2 = 16$ Finally, evaluate subtraction: $16 - 10 = 6$.

11.   9

Evaluate inside the parentheses first: $15 - 12 = 3$. Next, evaluate the exponent: $3^3 = 27$. Division is performed next: $27 \div 9 = 3$. Finally, $3 \times 3 = 9$.

12. 76

The term is already simplified in the parentheses, so evaluate the exponent to get $4^2 = 16$. Next, perform multiplication: $2 \times 16 = 32$. Finally, perform addition and subtraction from left to right: $(94 - 32) + 14 = 62 + 14 = 76$.

13. 2

The exponent is evaluated first: $10^2 = 100$. Next, perform division: $1{,}200 \div 100 = 12$. Next, multiply: $5 \times 2 = 10$. Finally, $12 - 10 = 2$.

14. commutative

15. associative

16. associative

17. both

18. 9

19. 18

20. 3.4

21. 12

CHAPTER 1 TEST

Now that you have studied the properties of numbers, try the following set of questions. Use the chapter material, including the practice questions throughout, to assist you in solving these problems. The answer explanations that follow will provide additional help.

1. Which set below features the even integers?

   (A) $\{2.4, 3.6, 4.8, 5.0, 6.2\}$

   (B) $\{...-6, -4, -2, 0, 2, 4, 6...\}$

   (C) $\{0, 1, 2, 3, 4...\}$

   (D) $\{...-7, -5, -3, -1, 1, 3, 5, 7...\}$

   (E) both A and B

2. All of the elements in the set $\frac{1}{4}, \frac{1}{3}, \frac{2}{3}, \frac{3}{4}, \frac{5}{6}$ are:

    (A) irrational numbers

    (B) integers

    (C) natural numbers

    (D) rational numbers

    (E) whole numbers

3. Simplify: $56 - 24 + 12$

    (A) 68   (B) 20   (C) 44   (D) 92   (E)   42

4. Simplify: $36 + 8 \div 2$

    (A) 40   (B) 22   (C) 52   (D) 25   (E)   88

5. Simplify: $32 - 4^2 \times 2$

    (A) 32   (B) 0   (C) 48   (D) 16   (E) 80

6. Simplify $(40 + 20) \div 10$

    (A) 3   (B) 42   (C) 5   (D) 1   (E) 6

7. Simplify: $48 \div 2 + 4 \times 4 - 2$

    (A) 110   (B) 3   (C) 38   (D) 30   (E)   56

8. The number sentence $5 \times 7 \times 20 = 5 \times 20 \times 7$ is an example of which property?

    (A) distributive   (B) commutative   (C) associative

    (D) order of operations   (E) none of the above

9. Which sentence below demonstrates the distributive property?

   (A) $8(40 + 7) = (8 + 40) \times (8 + 7)$

   (B) $8 + (40 \times 7) = (8 \times 40) + (8 \times 7)$

   (C) $8 (40 + 7) = (8 + 40) + 7$

   (D) $8(40 + 7) = (8 \times 40) + (8 \times 7)$

   (E) $8(40 + 7) = (40 + 7)8$

10. Choose from below to fill in the blank:
    $9(50 + 6) = (9 \times 50) + (9 \times \_\_\_)$

    (A) 56   (B) 6   (C) 9   (D) 15   (E) 54

11. Which example below demonstrates the associative property?

    (A) $(6.27 + 3.7) + 16.3 = 6.27 + (3.7 + 16.3)$

    (B) $6.27 + 3.7 + 16.3 = 16.3 + 6.27 + 3.7$

    (C) $7 \times (30 + 8) = (7 \times 30) + (7 \times 8)$

    (D) $(18 \times 3.7) + (18 \times 16.3) = 18(3.7 + 16.3)$

    (E) $18 \times (3.7 + 16.3) = (18 \times 3.7) + 16.3$

12. Choose the correct classification for the following numbers, in order: $-8, \frac{3}{4}, 0, \sqrt{2}, 9$

    (A) integer, irrational, whole number, irrational, integer

    (B) irrational, irrational, rational, irrational, rational

    (C) integer, rational, whole number, irrational, whole number

    (D) integer, irrational, whole number, irrational, rational

    (E) integer, rational, natural number, irrational, natural

13. Which of the following is an irrational number?

    (A) $3.77\overline{7}$ (B) $\sqrt{5}$ (C) $\frac{2}{3}$

    (D) both A and B (E) all of the above

14. The most specific name for the set of numbers {0, 1, 2, 3, 4...} is the
    (A) irrational numbers (B) natural numbers
    (C) whole numbers (D) rational numbers
    (E) real numbers

15. Simplify: $(15 - 12)^3 \div 9 \times 3$

    (A) 9 (B) 3 (C) $\frac{1}{3}$ (D) 1 (E) 27

16. Simplify: $10^2 \div (80 \div 4 \times 2)$

    (A) $\frac{2}{3}$ (B) 10 (C) $\frac{5}{2}$ (D) $\frac{1}{2}$ (E) 25

17. Simplify: $\dfrac{8(5 + 2) + 4}{5 + 5 \times 2}$

    (A) $\frac{19}{20}$ (B) 4 (C) 3 (D) $\frac{12}{5}$ (E) 5

18. Which choice shows the correct order of operations for the expression $12 - 2 \times 4 + 1$?
    (A) $12 - ((2 \times 4) + 1)$
    (B) $(12 - 2) \times (4 + 1)$
    (C) $((12 - 2) \times 4) + 1$
    (D) $12 - (2 \times (4 + 1))$
    (E) $(12 - (2 \times 4)) + 1$

19. Which choice shows the correct evaluation of the expression $8^2 - 4^2 \times 2$?

   (A) $(8 \times 2) - ((4 \times 2) \times 2)$

   (B) $((8 \times 2) - (4 \times 2)) \times 2$

   (C) $(8^2) - ((4^2) \times 2)$

   (D) $(64 - 16) \times 2$

   (E) $(8 \times 2) - 4 \times 2 \times 2$

20. Which statement illustrates the commutative property?

   (A) $19(70 + 7) = (19 \times 70) + (19 \times 7)$

   (B) $19 \times 70 \times 7 = 19 \times (70 \times 7)$

   (C) $19 + 3 + 21 = 19 + 21 + 3$

   (D) $19 \times 3 + 21 = 19 \times (3 + 21)$

   (E) $(19 \times 70) + (9 \times 70) = 70(19 + 9)$

21. Which of the following statements is true?

   (A) The commutative property allows you to change the grouping of addends in a sum.

   (B) The distributive property states that multiplication distributes over addition.

   (C) In the order of operations, multiplication is always evaluated before division.

   (D) all of the above

   (E) none of the above

22. Which set of numbers below contains at least one irrational number?
    (A) {... –5, –3, –1, 1, 3, 5...}

    (B) {0, 3, $\sqrt{16}$, $\frac{20}{4}$, 8}

    (C) {–4, 1, 5.101001000.., 8}

    (D) none of the above

    (E) both B and C

23. Which example below demonstrates the associative property of addition?
    (A) 5(6 + 19) = (5 + 6) + (5 + 19)
    (B) 74 + 14 + 86 = 74 x (14 + 86)
    (C) 74 + 14 + 86 = 14 + 86 + 74
    (D) 74 + 14 + 86 = 74 + (14 + 86)
    (E) 74 + 14 + 86 = (74 + 14) + 86

24. The first operation to be performed in the expression
    $$\frac{100 - \sqrt{32 \times 2} \div 2 + 4}{25}$$ is
    (A) square root (B) subtraction (C) division
    (D) addition (E) multiplication

25. Simplify $\dfrac{100 - \sqrt{32 \times 2} \div 2 + 4}{25}$

    (A) 4 (B) 2 (C) 96.25 (D) $\frac{7}{25}$ (E) 0.39

26. The area of a triangle is given by the formula
$A = (b \times h) \div 2$, where b is the length of the base and h is the height of the triangle. What is the area of a triangle with base of 5 cm and height of 8 cm?

(A) 40 cm^2 (B) 20 cm^2 (C) 160 cm^2

(D) 10 cm^2 (E) 200 cm^2

27. Centripetal force F of an object is $F = \dfrac{mv^2}{2}$,

where m is the mass and v is the velocity. What is the

centripetal force of an amusement ride of mass 5,000 grams

whose velocity is 5 m/sec?

(A) 3,125,000   (B) 25,000   (C) 12,500

(D) 62,500   (E) 6,250

28. The surface area of a cylinder is found using the formula
$2\pi r^2 + 2\pi rh$, where r is the radius of the base of the cylinder and h is the height of the cylinder. Find the surface area of a can with a radius of 3 inches and a height of 8 inches. Use 3.14 for the constant π.

(A) 3,956.4   (B) 207.24   (C) 66   (D) 60   (E) 56.52

Answers and Explanations

1. **B**

Choice (A) is not a set of even integers. Even though the numbers in this set have a last digit that is divisible by 2, they are not integers—they are decimals. Choice (C) is the set of whole numbers, both even and odd. Choice (D) is the set of odd integers.

2. **D**

A rational number is any number that can be expressed as a fraction. Choice (A) is incorrect; irrational numbers are real numbers that are NOT rational. Choice (B) does not describe the set—integers are not fractions. Choice (C) is not correct; natural numbers are numbers in

the set { 1, 2, 3, 4… }. Choice (E) is also incorrect; whole numbers
are numbers in the set { 0, 1, 2, 3, 4 … }.

3.   C

First subtract 24 from 56; 56 − 24 = 32. Now add this result to 12:
32 + 12 = 44.

4.   A

Evaluate division first: 8 ÷ 2 = 4. Then add: 36 + 4 = 40.

5.   B

The order of operations is exponents, then multiplication, and finally
subtraction. 4^2 =16, then 16 × 2 = 32. Finally, 32 − 32 = 0.

6.   E

Evaluate parentheses first, then divide. 20 + 40 = 60. 60 ÷ 10 = 6.

7.   C

Work from left to right, evaluating all division and multiplication first.
Thus, 48 ÷ 2 = 24. Next, 4 × 4 = 16. Then, 24 + 16 = 40. Finally,
40 − 2 = 38.

8.   B

This example shows that changing the order of the factors does not
change the resulting product, which is by definition the commutative
property. Choice (A), the distributive property, states that multiplication distributes over addition. Choice (C), the associative property,
deals with the grouping of factors, not the order. Choice (D), order of
operations, is a ruling on the precedence of operations.

9.   D

Choice (D) shows that multiplication distributes over addition.
Choices (A), (B), and (C) are all false statements; they are not examples of any property of numbers. If you simplify the equation in
choice (A) by using the order of operations, you get 8 × 47 = 48 ×
15, or 376 = 720, which is false. If you simplify the equation in
choice (B) by using the order of operations, you get 8 + 280 = 320
+ 56, or 288 = 376, which is false. If you simplify the equation in
choice (C) by using the order of operations, you get 8 × 47 = 48 +
7, or 376 = 55, which is false. Choice (E) shows an example of the

commutative property, which states that you can change the order of factors without changing the product.

10. B

This example of the distributive property is missing the 6, which is the second part of the sum that 9 distributes over.

11. A

Choice (A) indicates that changing the grouping of addends does not affect the resultant sum. This is, by definition, the associative property. Choice (B) is a demonstration of the commutative property. Choices (C) and (D) are examples of the distributive property. If you simplify the equation in choice (E) by using the order of operations, you get $18 \times 20 = 66.6 + 16.3$, or $360 = 82.9$, which is false.

12. C

Choice (C) is the correct classification for each of the numbers listed. Negative eight is an integer, an even integer, and a rational number. Three fourths is a rational number. Zero is an integer, a whole number, and a rational number. The square root of 2 is irrational. Nine is an integer, a whole number, a natural number, and a rational number.

13. B

Choice (B) is the only irrational number. A rational number is a number that can be expressed as a fraction. Choice (A), $3.7\overline{77} = 3\frac{7}{9} = \frac{34}{9}$ and thus is a rational number. Choice (C), a fraction, is by definition a rational number. Choice (B) is the only choice that cannot be expressed as a fraction. When converted to a decimal equivalent, $\sqrt{5}$ neither terminates nor repeats.

14. C

The most specific name for this set is the whole numbers, choice (C). This set of numbers is a set of rational numbers—all can be expressed as fractions. Therefore, choice (A) is false. Choice (D) is a true classification for the set, but it is not the most *specific* name for the set. Choice (B) is not a correct classification; the natural numbers do not include the number 0. Choice (E) is true—the numbers in this set are all real numbers, but it is not the most *specific* classification.

15. A

Evaluate parentheses first: $15 - 12 = 3$. Next, evaluate the exponent: $3^3 = 3 \times 3 \times 3 = 27$. Next, divide: $27 \div 9 = 3$. Finally, multiply: $3 \times 3 = 9$.

16. C

First, simplify within the parentheses. Division is located to the left of multiplication, so divide first: $80 \div 4 = 20$. Now, multiply 20 times 2 to get 40. The expression within parentheses is simplified, so evaluate the exponent next. $10^2 = 100$. The final step is to divide 100 by 40. $\frac{100}{40} \div \frac{20}{20} = \frac{5}{2}$.

17. B

This is a fractional expression. The fraction bar acts as a grouping symbol, so evaluate the numerator, evaluate the denominator, and then divide. In the numerator, add $5 + 2 = 7$ because this is contained within parentheses. Next, multiply by 8: $7 \times 8 = 56$. Add 4: $56 + 4 = 60$. In the denominator, multiplication is evaluated before addition. $5 \times 2 = 10$, and then $5 + 10 = 15$. Finally, $60 \div 15 = 4$.

18. E

Parentheses have been added to the answer choices to show the order of preference for the operations. Multiplication is performed first, followed by subtraction (it is to the left of the addition operation), and then addition.

19. C

Evaluate the exponents first. Choice (C) shows this, because these exponents are enclosed in parentheses. The outer parentheses surrounding $4^2 \times 2$ indicate that multiplication is evaluated before subtraction.

20. C

Choice (C) is the only example that demonstrates the commutative property, which states that changing the order of addends does not change the resultant sum. Choices (A) and (E) are examples of the distributive property. Choice (B) is a demonstration of the associative property. Choice (D) is not an example of any property of numbers; it is false. If you simplify the equation in choice (D) by using the order of operations, you get $57 + 21 = 19 \times 24$, or $78 = 456$.

21. B

Choice (B) is the only true statement—multiplication distributes over addition. Choice (A) is not true—the commutative property allows you to change the order, not the grouping, of addends in a sum. Choice (C) is not always true. In the order of operations, multiplication is performed before division if it is to the left of division. Otherwise, division is performed first.

22. C

Choice (C) has one element that is irrational—5.101001000... This decimal number neither terminates nor repeats. Notice that in choice (B), you may have been tempted to choose $\sqrt{16}$ as irrational. But this expression simplifies to 4, which is a rational number. Not all radicals are irrational numbers.

23. D

This example shows that you can change the grouping of the addends without changing the resultant sum. Be aware that in choice (E), even though parentheses were inserted, the grouping did not change; order of operations makes that first set of parentheses understood.

24. E

The first operation performed will be multiplication. This is imbedded in the radical sign, and a radical sign is like a grouping symbol—it is first priority for this problem. Choice (A), the square root, would be performed second. Choice (C), division, would be performed third, after the square root operation. Division takes precedence over addition or subtraction. Choice (D), addition, would be the last operation performed on the numerator. Subtraction would be first, because addition and subtraction are evaluated from left to right. The division indicated by the fraction bar would be performed last, because a fraction bar is a grouping symbol.

25. A

As described in problem 25, multiplication will be first, and then the square root function: $32 \times 2 = 64$; $\sqrt{64} = 8$. The expression is now $\frac{100 - 8 \div 2 + 4}{25}$. Division is performed next: $8 \div 2 = 4$. The numerator is now $100 - 4 + 4$, or just 100. Finally, $100 \div 25 = 4$.

26. B

Multiply base times height and then divide by 2: $8 \times 5 = 40$; $40 \div 2 = 20$ cm^2.

27. D

This is a formula, and you must apply the correct order of operations to get the correct answer. Substitute in the correct values for the mass and velocity: $F = \frac{5,000(5^2)}{2}$. First, evaluate 5^2 to get 25, and then multiply by 5,000. The final step is to divide by 2: $(5,000 \times 25) \div 2 = 125,000 \div 2 = 62,500$.

28. B

Substitute in the correct values for π (3.14), r (the radius) and h (the height): $2 \times 3.14 \times 32 + 2 \times 3.14 \times 3 \times 8$. There are no parentheses, so evaluate 32 first to get:

$2 \times 3.14 \times 9 + 2 \times 3.14 \times 3 \times 8$.

Next, multiply from left to right:

$((2 \times 3.14) \times 9) + ((2 \times 3.14) \times 3 \times 8) =$

$(6.28 \times 9) + ((6.28 \times 3) \times 8) =$

$56.52 + (18.84 \times 8) =$

$56.52 + 150.72 =$

$207.24.$

CHAPTER 2

Integers

BUILDING BLOCK QUIZ

The Building Block Quiz for chapter 2 includes some questions to help you review the concepts from chapter 1, Properties of Numbers. You will also find questions to help get you ready for this chapter on integers. Take this quiz first before starting the chapter to help lay the foundation for mastery and understanding.

1. Simplify the expression: $(7-3)^2 + 12 \times 2$
   (A) 28   (B) 32   (C) 40   (D) 56   (E) 124

2. Which of the following is an example of the commutative property of addition?
   (A) $3 + (4+5) = (3+4) + 5$
   (B) $6(2+9) = 6 \times 2 + 6 \times 9$
   (C) $14 + -14 = 0$
   (D) $5 + 18 = 18 + 5$
   (E) $99 + 0 = 99$

3. Which of the following is an example of an irrational number?
   (A) 0.5   (B) $\dfrac{3}{4}$ (C) $\sqrt{2}$ (D) $0.\overline{3}$ (E) 14

4. The integer 18 is located how many spaces to the right of 0 on a number line?
   (A) 0   (B) 8   (C) –8   (D) 18   (E) –18

5. Simplify the following: $-10 + (-3) \times (-2) - 9$

   (A) -25 (B) -13 (C) 7 (D) 17 (E) 143

6. Which of the following is a number that has the same absolute value as 6?

   (A) 0 (B) 1 (C) -1 (D) -6 (E) 36

7 $-12 \times (-1) \times 3 =$

   (A) -36 (B) -15 (C) -10 (D) 15 (E) 36

8. For the inequality $-22 <$ _____ < -13, which of the following could be placed in the blank to make the statement true?

   (A) -23 (B) -12 (C) -21 (D) -42 (E) 23

9. Evaluate the expression $|-3-4|$.

   (A) -12 (B) -7 (C) -1 (D) 1 (E) 7

10. The low temperature of a certain city was $-3°$ on Monday. It then dropped 9 degrees on Tuesday and was 15 degrees higher than Tuesday on Wednesday. What was the low temperature on Wednesday?

    (A) -12 (B) -3 (C) -2 (D) 2 (E) 3

Answers and Explanations

1.   C

This problem reviews the concept of order of operations. Use the correct order of operations to simplify. First, evaluate within the parentheses. Since $7 - 3 = 4$, the expression becomes $4^2 + 12 \times 2$. Next, evaluate the exponent to get $16 + 12 \times 2$. Since multiplication comes before addition, multiply 12 and 2 to get 24. The expression is now $16 + 24$, which is equal to 40.

2.   D

This problem reviews the concept of the commutative property of addition, which changes the order of the numbers being added, but does not change the solution: $5 + 18 = 23$ and also $18 + 5 = 23$. Choice (A) is an example of the associative property of addition, where the grouping of the numbers changed. Choice (B) is an example of the distributive property. Choice (C) is incorrect because it is the inverse property of addition. Choice (E) is an example of the identity property of addition.

3.   C

This problem reviews the different types of numbers. An irrational number is a number that cannot be expressed as a repeating or terminating decimal. Since $\sqrt{2}$ is a non-repeating, non-terminating decimal, it is an example of an irrational number. Choice (A) is a terminating decimal; choice (B) is equivalent to 0.75, which is also a terminating decimal; choice (D) is a repeating decimal; and choice (E) is a whole number, which is also a terminating decimal.

4.   D

This question tests the concept of working on a real number line. The integer 18 is located 18 places away from 0 on a number line. Remember that a measure of distance is always positive, no matter what direction you are moving.

5.   B

This question tests the concept of using order of operations with integers. Using order of operations, multiply -3×-2 first to get 6. The expression then becomes $-10 + 6 - 9$. Combine the values in order from left to right.

$$-10 + 6 = -4$$
$$-4 - 9 = -4 + -9$$
$$= -13.$$

6.   D

Since 6 and -6 are both 6 units away from 0 on a number line, they each have an absolute value of 6. Choice (A) is not the answer because the absolute value of 0 is 0; choices (B) and (C) each have an absolute value of 1, and choice (E) has an absolute value of 36.

7. E

To find the product of $-12 \times -1 \times 3$, first find the product of -12 and -1, which is equal to $+12$. Then find the product of 12×3, which is equal to 36. Since there is an even number of negatives in the problem, the answer is positive.

8. C

This problem tests the concept of ordering integers. To solve this problem, you need to find a number that is between -22 and -13 on a number line. Since -21 is larger than -22 and smaller than -13, it is between those two values.

9. E

This question tests evaluating absolute value expressions. To simplify the expression $|-3-4|$, first evaluate inside the absolute value bars: $-3 - 4 = -3 + -4 = -7$. The problem now becomes $|-7|$. Since -7 is 7 units away from 0 on a number line, $|-7| = 7$.

10. E

This question tests the concept of solving word problems using integers. The temperatures in this problem can be translated into the expression $-3 - 9 + 15$. To simplify, $-3 - 9 = -3 + -9 = -12$. Finally, $-12 + 15 = 3$.

WORKING WITH INTEGERS

This chapter will focus on the set of numbers known as **integers**. You encounter integers in many different places each day. They are used to express temperatures above and below 0, a loss or gain of yards when playing certain sports, and the highs and lows of the stock market, to name a few. Integers provide a foundation for the real number system. To help you in your study, this chapter will provide information and practice with absolute value and ordering integers, and operations using both.

> **MATH SPEAK**
>
> **Integers** are the set of whole numbers and their opposites. As a set, the integers are written as $\{\ldots, -3, -2, -1, 0, 1, 2, 3, \ldots\}$.

Each integer has a location on a real **number line**, where the sign of the number determines to which side of 0 the number is located. For example, positive 5, which can also be written as +5 or simply 5, is located 5 units to the right of 0, as shown in the diagram below.

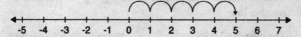

Negative 8, which can also be written as –8, is located 8 units to the left of 0, as shown in the diagram below.

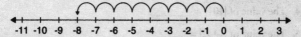

It is important to remember that the number 0 is neither positive nor negative.

Absolute Value

Before discussing operations with integers, it is helpful to understand the concept of absolute value.

> **MATH SPEAK**
>
> **Absolute value** is the number of units a number is away from 0 on a number line.

Since absolute value is a measure of distance, it is always a positive value. The symbol for absolute value is two bars on either side of a numerical value or expression. For example, the absolute value of 4 is written as |4|, and since it is 4 units away from 0 on a number line, |4| = 4. The absolute value of −7 is written as |−7|, and since −7 is 7 units away from 0 on a number line, |−7| = 7.

For practice with integers and absolute value, try the set of exercises below.

Practice 1

Fill in the blank with the best possible word. Answers and explanations are located at the end of the chapter.

1. Negative numbers are located on the _____ side of 0 on a number line.

2. The integer −2 is located _____ spaces to the left of 0 on a number line.

3. The absolute value of 5 is equal to _____.

4. The value of |101| = _____.

5. The value of |−442| = _____.

6. The integer _____ is neither positive nor negative.

More information on working with absolute value will follow later in this chapter.

Ordering Integers

The value of an integer is determined by its location on the real number line, where negative numbers appear to the left of 0 and positive numbers are located to the right of 0. When comparing integers, first determine their locations on the number line. A number farther to the right will be larger in value than a number farther to the left. For example, when comparing −8 and −9, −9 is farther to the left on a number line—therefore, −9 is less than −8. This concept can also be written using the symbol for less than, and would appear as −9 < −8.

For practice with ordering, try the following exercise.

Place the following integers in ascending order: 1, –10, –1, –100.

Ascending order means arranging the numbers from smallest to largest. Start with the value that is the *farthest* to the left on a number line, which would be –100. This is the smallest value in the list. The next smallest is –10, and then –1. The only positive number in the list, 1, is the largest number. The integers listed in ascending order would be –100 < –10 < –1 < 1.

Practice 2

Now, try the following true/false questions to test your ordering skills. Answers and explanations are located at the end of the chapter.

7. **T F** The integer –5 is larger than the integer –6.

8. **T F** The set –65, –60, –59 is correctly listed in ascending order.

9. **T F** –11 < –12

INTEGER OPERATIONS

Adding Integers

When adding integers, the sign of the numbers involved is very important. In order to help you visualize adding integers, think of any positive integer as a group of that many positives, and any negative integer as a group of that many negatives. Thus, +5 would be represented as 5 positives.

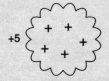

In the same way, −7 would be represented as 7 negatives.

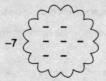

Now, keep in mind that any time one positive and one negative are grouped together, they cancel each other out, making a neutral. You can consider this as +1 + −1 = 0.

$$\boxed{+ \ -} \text{ equals a neutral}$$

When adding integers that have the same sign, just add the absolute values of the numbers and keep the sign.

For example, +6 + +7 = + 13

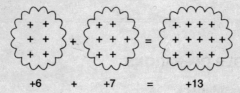

and −4 + −8 = −12

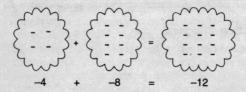

When adding integers that have different signs, recall that each negative sign can pair with a positive sign to form a neutral. Each member of the pair cancels the other out. Therefore, the solution to an addition problem with integers of different signs will be the remaining positives or negatives that did not form a pair. For example, in the problem −3 + 5, 3 negatives pair with 3 of the 5 positives to form 3 neutral pairs.

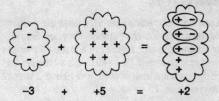

You now have 2 positives left over that did not form a pair. Therefore, the solution to the problem −3 + 5 is 2.

A general rule for adding integers with different signs is to subtract their absolute values and keep the sign of the number with the larger absolute value as your answer. In the example 18 + −25, subtract the absolute values to get 25 − 18 = 7. Since −25 has a larger absolute value, take the negative sign for your answer. The final answer is −7.

REMEMBER THIS!

When adding integers:

1. If the signs are the same: Add, and keep the sign.

2. If the signs are different: Subtract, and take the sign of the number with the larger absolute value.

Subtracting Integers

The subtraction of any two integers can also be expressed as adding the opposite of the number being subtracted. This way, the concept can be simplified into an addition problem, and you only have two rules to commit to memory.

Here are a few examples to demonstrate how this works.

Find the value of $19 - (-2)$.

Since this is a problem subtracting integers, change the problem so that it is adding the opposite of the number subtracted. So the subtraction sign changes to an addition sign and the -2 changes to a $+2$. The problem now becomes $19 + (+2)$. Follow the rules for addition. Since the signs are both positive, add 19 and 2 to get 21 and keep the solution positive. $19 - (-2) = 21$.

Find the value of $-45 - 9$.

As in the previous example, change the subtraction sign to an addition sign and change the $+9$ to a -9. The problem now becomes $-45 + (-9)$. Since the signs are both negative, add the absolute values to get $45 + 9 = 54$ and keep the solution negative. $-45 - 9 = -54$.

The temperature on a certain day dropped from $-4°$ F to $-17°$ F. What is the difference in temperature for that day?

This example illustrates how negative integers can be used to show temperature. Since you are looking for the difference between the two temperatures, subtract the two values. $-4 - (-17)$ then becomes $-4 + 17$ after the subtraction sign is changed to addition and the sign of -17 is changed to a positive. Since the signs are now different, subtract the absolute values and take the sign of the larger absolute value. $17 - 4 = 13$. The difference in temperature is 13 degrees.

FLASHBACK

Use the number properties studied in chapter 1 to help simplify expressions with integers. For example, take the expression $-4 + 5 + -6$. You can use the commutative property of addition to change the order of the expression to $-4 + -6 + 5$. Now the negative numbers are together and can easily be combined. Since $-4 + -6 = -10$, now add $-10 + 5$ to get a final answer of -5.

Practice 3

Try the following set of questions to practice adding and subtracting integers. For each numbered problem in the first column, find the letter that corresponds with the correct sum or difference. Answers are located at the end of the chapter.

10. $-12 + 4$	A. -32
11. $14 - (-5)$	B -1
12. $-201 - (-200)$	C. 18
13. $-18 + -14$	D. 19
14. $23 + -5$	E. -8

Multiplying and Dividing Integers

Multiplication and division of integers is a bit more straightforward than adding and subtracting. Regardless of the numbers' signs, you multiply or divide the absolute values of the numbers just as you would when you first learned how to multiply and divide. The only question is whether the solution is positive or negative. Since multiplication is repeated addition, use the principles of addition to make sense of the rules. In the example $6 \times (-2)$, this is the same as adding 6 groups of -2. Therefore, the answer would be -12.

For the problem $(-5) \times (-3)$, this is the same as the opposite of adding 5 groups of -3. This would have a result of 15. There are two negatives in the problem, so they cancel each other out.

Since division follows the same principle, a problem such as $-12 \div 4$ would result in an answer of -3. There is only one negative in the problem, so the answer will also be negative. The negative has nothing to cancel out with.

In the example $-4 \times -9 \times -2$, the result will be the opposite of the product of 4, 9, and 2, which is -72. The solution here is negative because there are two negatives that pair up to cancel out and one negative sign left over.

> **REMEMBER THIS!**
>
> When multiplying and dividing integers, if there is an even number of negatives in the problem, the solution will be positive. If there is an odd number of negatives, the solution will be negative.

Practice 4

Remember, repetition is the key to mastery! Use the following bank of integers to answer the questions below on multiplying and dividing integers. Answers and explanations are located at the end of the chapter.

9	−30	−12	−9	40	−20
21	30	−35	−40	−1	−3

15. $-5\,(-6) =$

16. $-5 \times 6 =$

17. $-36 \div -4 =$

18. $21 \div -7 =$

19. $-4 \times -2 \times -5$

Working with Operations and Absolute Value

When working with absolute value, expressions are simplified much in the same way as expressions that contain parentheses. Any operations contained within the absolute value bars are evaluated first, using the correct order of operations. Take, for example, the expression $|-16 + 7|$. The first step in simplifying is to combine -16 and 7, which is equal to -9. Remember that the -9 is still inside the absolute value bars. Now, evaluate the absolute value of -9: $|-9| = 9$

In a contrasting example such as $|-16| + |7|$, only single values are contained within the absolute value bars. Evaluate the absolute values first. Since $|-16| = 16$ and $|7| = 7$, the problem then becomes $16 + 7 = 23$. The key for both of these examples is to simplify any absolute value problem down to a single value within the absolute value bars first, and then calculate the absolute value of the number. Refer to chapter 1 of this book for any questions on order of operations.

Practice 5

Try this practice set to help you assess your understanding of operations and absolute value.

20. Find the value of the expression $|24 - (-4)|$.

21. Find the value of the expression $|-8| + |6|$.

22. Find the value of the expression $-|-16| - |5|$.

23. Is there a difference in the solutions of the two expressions $|-15 + 9|$ and $|-15| + |9|$? Explain your answer.

SUMMARY

To review, here are the main points about integers and absolute value that you should take from this chapter:

- The absolute value of an integer is the distance the number is away from 0 on a number line. This value is always positive.

- To compare integers, the number farther to the left on a number line is the smaller integer.

- When adding integers with like signs, add the absolute values and keep the sign. If the signs are different, subtract the absolute values and take the sign of the number with the larger absolute value.

- Any subtraction problem can be rewritten as an addition problem by changing the subtraction sign to an addition sign and switching the sign of the number being subtracted. Then simply follow the rules for addition.

- When multiplying and dividing integers, an even number of negatives results in a positive answer. An odd number of negatives results in a negative answer.
- Evaluate within any absolute value bars first in a question, and then take the absolute value when a single value appears within the bars.

Practice Answers and Explanations

1. left

On any horizontal number line, negative numbers appear to the left of 0. The value of the numbers decreases the farther to the left you move.

2. 2

The integer −2 is located 2 places away from 0 on a number line. Since distance is always a positive value, the answer is positive 2 even though you moved to the left to get to −2.

3. 5

The absolute value of a number is the distance the number is away from 0 on a number line. The number 5 is 5 units away from 0.

4. 101

The expression |101| means the absolute value of positive 101, which is equal to 101. The integer 101 is 101 units away from 0 on a number line.

5. 442

The expression |−442| means the absolute value of negative 442, which is equal to 442. The integer −442 is 442 units away from 0 on a number line.

6. 0

The integer 0 is considered neutral; it is neither a positive nor negative integer.

7. True

Negative 5 is to the right of −6 on a number line.

8. True

These numbers are in order from smallest to largest.

9. False

Negative 11 is greater than −12 because it is closer to 0 on a number line.

10. E

11. D

12. B

13. A

14. C

15. 30

The solution is positive because there are two negatives.

16. −30

The solution is negative because there is an odd number of negatives.

17. 9

The solution is positive because there are two negatives.

18. −3

The solution is negative because there is an odd number of negatives.

19. −40

The solution is negative because there is an odd number of negatives.

20. 28

$|28| = 28$

21. 14

$8 + 6 = 14$

22. −21

$-|-16|-|5| = -16 - 5 = -16 + -5 = -21$

23. Yes

When evaluating the first expression, combine −15 and 9 first to get |−6| = 6. In the second expression, evaluate the absolute values first to get 15 + 9, which is equal to 24. The order of operations for the two expressions was different, resulting in two different solutions.

CHAPTER 2 TEST

Try the following questions to test your knowledge of integers. Use the lessons and practice from the chapter to answer any questions you may have. The answer explanations following this section will also provide help and guidance.

1. The number −12 is how many units away from 0 on a number line?

   (A) −13   (B) −12   (C) −1   (D) 1   (E) 12

2. The absolute value of an integer is always

   (A) equal to zero   (B) negative

   (C) positive or equal to zero   (D) either positive or negative

   (E) equal to the opposite of the number

3. The value of = $|-98|$ _____?

   (A) −99   (B) −98   (C) −97   (D) 98   (E) 99

4. The value of $|23|$ = _____?

   (A) −24   (B) −23   (C) 23   (D) 24   (E) 46

5. Which of the following answer choices are NOT in ascending order?

   (A) 0, 2, 4   (B) −5, −7, −8   (C) −21, −20, −19

   (D) −90, −80, −70   (E) none of these

6. For the inequality $-35 < \underline{\hphantom{xxx}} < -24$, which of the following could be placed in the blank to make the inequality true?

   (A) -46 (B) -36 (C) -34 (D) -23 (E) -20

7. Evaluate: $-13 + -6$

   (A) -19 (B) -7 (C) 7 (D) 18 (E) 19

8. Evaluate: $20 + (-5)$

   (A) -25 (B) -15 (C) -4 (D) 15 (E) 25

9. Evaluate: $-19 + 4 + (-2)$

   (A) -25 (B) -21 (C) -17 (D) -13 (E) 13

10. Evaluate: $-27 - 11$

    (A) -38 (B) -37 (C) -16 (D) 37 (E) 38

11. Evaluate: $-56 - (-9)$

    (A) 65 (B) 47 (C) -47 (D) -65 (E) 504

12. Evaluate: $-18 \div (-2)$

    (A) 9 (B) 6 (C) -9 (D) -20 (E) -36

13. Evaluate: -77×2

    (A) -154 (B) 154 (C) -75 (D) -79 (E) 79

14. Evaluate: $-4 \times -1 \times -9$

    (A) -14 (B) 14 (C) -36 (D) 36 (E) -32

15. Evaluate: $-6 + (14 - (-2)) - 3^2$

    (A) 1 (B) 4 (C) 13 (D) -3 (E) 0

16. Find the value of the expression $|4| - |5|$.

    (A) -20 (B) -9 (C) -1 (D) 1 (E) 9

17. Find the value of the expression $|-7 + -78|$.

    (A) -85 (B) -71 (C) 85 (D) -86 (E) 71

18. Find the value of the expression $-|-10 \times 5 + (-12)|$

    (A) -38 (B) -62 (C) -70 (D) 62 (E) 38

19. While performing an experiment, a science student measured
    a substance to have a temperature of $-14°F$. She then raised
    the temperature of the substance 15 degrees, before lowering
    the temperature 18 degrees. What was the final temperature
    of the substance after the experiment?

    (A) $19°F$ (B) $15°F$ (C) $-13°F$ (D) $-17°F$ (E) $-18°F$

20. Marty checks her bank account and finds that she has a
    balance of $231.28. She then writes checks in the amounts of
    $64.75, $122.20, and $49.08 to pay some of her bills. What is
    the balance in her account after paying the bills?

    (A) $-\$236.03$ (B) $-\$4.75$ (C) $\$4.75$

    (D) $\$236.03$ (E) $\$467.31$

21. Starting from the second floor, a person takes an elevator
    down 1 floor, up 14 floors, down 6 floors, and then down
    another 2 floors. On what floor does this person end up?

    (A) 2nd   (B) 5th   (C) 7th   (D) 21st   (E) 23rd

Answers and Explanations

1.   **E**

The number -12 is 12 units to the left of 0 on a number line. Choices
(A), (B), and (C) cannot be correct because distance cannot be a
negative value. Choice (E) is not the answer because only the numbers
-1 and 1 are one unit from 0 on a number line.

2. C

The absolute value of a number is the distance the number is away from 0 on a number line. Since it is a measure of distance, it is always positive or equal to zero. Choice (A) is not the correct answer because only the number 0 has an absolute value of 0. Choices (B) and (D) are not correct because the absolute value of a number is never negative. Choice (E) is not always correct. For example, the absolute value of 3 is 3, not −3.

3. D

The expression |−98| represents the absolute value of −98. Since −98 is 98 units to the left of 0 on a number line, the absolute value of −98 is equal to 98.

4. C

The expression |23| means the absolute value of positive 23, which is equal to 23. The integer 23 is 23 units away from 0 on a number line.

5. B

Each of the answer choices is listed in order from smallest to largest except for the numbers in choice (B). Since −5 is greater than −7, and −7 is greater than −8, these numbers are listed in descending, or decreasing, order.

6. C

To solve this problem, you need to find a number that is between −35 and −24 on a number line. Since −34 is larger than −35 and smaller than −24, it is between the two values.

7. A

Since you are adding and the signs of the numbers are the same, add the absolute values and keep the sign. −13 + −6 = −19.

8. D

Since you are adding and the signs are different, subtract the absolute values of the numbers and keep the sign of the larger absolute value for your answer. 20 − 5 = 15; the sign of the larger is positive so the final answer is +15.

9. C

Since you are adding and the signs of the first two numbers are different, subtract the absolute values and keep the sign of the number with the larger absolute value. $-19 + 4 = -15$. Now add $-15 + -2$ by adding the absolute values and keeping the negative sign. $-15 + -2 = -17$.

10. A

Since you are subtracting, change the subtraction to addition and the sign of the number being subtracted to its opposite. $-27 - 11$ becomes $-27 + -11$. The operation is now addition and the signs are the same, so add the absolute values and keep the sign. $-27 + -11 = -38$.

11. C

Since you are subtracting, change the subtraction to addition and the sign of the number being subtracted to its opposite. $-56 - (-9)$ becomes $-56 + 9$. The operation is now addition and the signs are different, so subtract the absolute values and take the sign of the number with the larger absolute value for the answer. $56 - 9 = 47$; the sign of the larger is negative. Thus, the answer is -47.

12. A

Divide 18 by 2 to get 9. Since there is an even number of negatives in the problem, the answer is $+9$.

13. A

Multiply 77 by 2 to get 154. Since there is an odd number of negatives in the problem, the final answer is -154.

14. C

Multiply $4 \times 1 \times 9$ to get 36. Since there is an odd number of negatives in the problem, the final answer is -36.

15. A

Evaluate this expression by using the order of operations. The first step is to evaluate within the parentheses. In the expression $14 - (-2)$, change the subtraction sign to addition and the sign on -2 to $+2$. It then becomes $14 + 2 = 16$. Then, evaluate the exponent of 2 on the base of 3. $3^2 = 9$. The entire expression is now $-6 + 16 - 9$. Combine -6 and 16 to get 10 by subtracting the absolute values and making the result positive. To complete the problem subtract 10 − 9 to get a final answer of 1.

16. C

Since each of the terms inside the absolute value bars is a single value, evaluate the absolute values first. |4| − |5| becomes 4 −5, which is equal to −1.

17. C

Evaluate within the absolute value bars first. |−7 + − 78| becomes |−85| because the signs on the values are the same and you are adding. Since the absolute value of −85 is 85, the final answer is 85.

18. B

Start by evaluating within the absolute value bars by multiplying −10 by 5. The expression then becomes −|−50 + (−12)|. Continue to simplify within the bars by adding −50 and −12 to get −62. Now the expression is −|−62|. Evaluate the absolute value of −62 to get 62. The solution includes the negative in front of the absolute value bars in the original question, so the final answer is −62.

19. D

This problem can be solved by using the expression −14 + 15 + −18, since the substance started out at −14°, was raised 15°, and then was lowered 18°. Since −14 + 15 = 1, and 1 + −18 = −17, the final answer is −17° F.

20. B

Since Marty has a starting balance of $231.28, take that amount and subtract the amount of each check. 231.28 − (64.75 + 122.20 + 49.08) = 231.28 − 236.03 = 231.28 + −236.03 = −4.75. Marty has a negative balance of $4.75.

21. C

Write an expression for the elevator trip that starts at the 2nd floor. 2 − 1 + 14 − 6 − 2. Changing each subtraction sign to addition the expression becomes 2 + −1 + 14 + −6 + −2. Use the commutative property from chapter 1 to change the order of the expression to 2 + 14 + −1 + −6 + −2. Combining the negative values and the positive values yields 16 + −9 = 7. The person ended up on the seventh floor.

CHAPTER 3

Fractions and Decimals

BUILDING BLOCK QUIZ

This chapter will introduce operations with fractions and decimals. Start off your study with this Building Block Quiz. Remember that the first three problems will review integer operations from chapter 2.

1. $-19 - 5 =$
    (A) 26   (B) 24   (C) -24 (D) -14 (E) 14

2. $|36 - 45| =$
    (A) -9 (B) 9   (C) -81 (D) 81   (E) 11

3. $\dfrac{-56}{-8} =$

    (A) 7   (B) -7 (C) 9   (D) -9 (E) -64

4. Which of the following is equivalent to $\dfrac{28}{36}$?

    (A) $\dfrac{4}{9}$ (B) $\dfrac{9}{4}$ (C) $1\dfrac{7}{9}$ (D) $\dfrac{4}{6}$ (E) $\dfrac{7}{9}$

5. What is the least common multiple of 15 and 35?
    (A) 75   (B) 5   (C) 3   (D) 105   (E) 525

6. Which of the following is equivalent to 1.75?

    (A) $1\frac{7}{5}$ (B) $\frac{17}{5}$ (C) $1\frac{3}{4}$ (D) $\frac{175}{10}$ (E) $\frac{7}{5}$

7. $2.035 + 3.9 =$
    (A) 2.074   (B) 5.935   (C) .5395   (D) .2074   (E) 2.064

8. Which of the following shows numbers arranged in increasing order?

    (A) $-\frac{3}{4}, \frac{2}{5}, 0.39, \frac{1}{2}$

    (B) $\frac{2}{5}, \frac{1}{2}, -\frac{3}{4}, 0.39$

    (C) $-\frac{3}{4}, 0.39, \frac{2}{5}, \frac{1}{2}$

    (D) $-\frac{3}{4}, \frac{1}{2}, 0.39, \frac{2}{5}$

    (E) $\frac{1}{2}, 0.39, \frac{2}{5}, -\frac{3}{4}$

9. $\dfrac{\frac{2}{3}}{8} =$

    (A) $\frac{1}{12}$ (B) $\frac{16}{3}$ (C) $\frac{3}{16}$ (D) $\frac{6}{8}$ (E) $2\frac{3}{8}$

10. The owner of Mr. Tasty Cone decided to survey his customers one day on their favorite ice cream flavors. If $\frac{1}{3}$ of the people polled preferred chocolate ice cream, and $\frac{2}{5}$ of the people polled preferred vanilla ice cream, what fraction of the people preferred either chocolate or vanilla?

(A) $\frac{3}{8}$ (B) $\frac{5}{8}$ (C) $\frac{11}{15}$ (D) $\frac{4}{15}$ (E) $\frac{3}{15}$

Answers and Explanations

1. C

This problem reviews integer subtraction. When you subtract, change the problem to add the opposite of the number being subtracted. For example, $-19 - 5$ becomes $-19 + -5$. When you add two numbers with the same sign, add the numbers and keep the sign. $-19 + -5 = -24$.

2. B

This problem tests your knowledge of absolute value, as well as integer subtraction. First, simplify inside the absolute value sign. $36 - 45$ is the same as $36 + -45$. When you add two numbers with different signs, subtract the numbers as if they were positive and take the sign of the number with the larger absolute value. $45 - 36 = 9$, and the sign so far is negative. Because -9 is enclosed in the absolute value symbols, the answer is 9.

3. A

This problem reviews integer division. A negative divided by a negative is a positive. $-56 \div -8 = 7$.

4. E

This problem tests your knowledge of equivalent fractions. To simplify, divide the numerator (the top number) and the denominator (the bottom number) by the greatest common factor of 4. $\frac{28}{36} \div \frac{4}{4} = \frac{7}{9}$.

5. D

This problem deals with the least common multiple of two numbers. The prime factorization of 15 is 3×5, and the prime factorization of 35 is 5×7. The factor of 5 is common to both, so the least common multiple is $3 \times 5 \times 7 = 105$. Choice (A) is a multiple of 15, but not of 35. Choice (B) is the greatest common factor of 15 and 35. Choice (C) is a factor of 15, but not of 35. Choice (E) is a multiple of both 15 and 35 ($15 \times 35 = 525$), but it is not the least common multiple.

6. C

This problem tests conversions between fractions and decimals. 1.75 is equal to $\frac{175}{100}$. Divide the numerator and the denominator by the greatest common factor of 25 to simplify: $\frac{175}{100} \div \frac{25}{25} = \frac{7}{4}$, or $1\frac{3}{4}$. Choice (A) is $1 + (7 + 5) = 1 + 1.4 = 2.4$. Choice (B) is $17 + 5 = 3.4$. Choice (D) is $175 + 10$, or 17.5. Choice (E) is $7 + 5 = 1.4$.

7. B

This problem tests decimal addition. Remember, it is important to line up decimal points when adding and subtracting decimals. Tack two zeroes onto the end of 3.9 to get 3.900 and then add. $2.035 + 3.900 = 5.935$.

8. C

This question tests both your knowledge of fraction to decimal conversions and your knowledge of ordering decimals. Because the sets have both fractions and decimal elements, it is advisable to first convert all fractions to decimals, making it easier to check for increasing order: $-\frac{3}{4} = -0.75$, $\frac{2}{5} = 0.4$, $\frac{1}{2} = \frac{5}{10} = 0.5$. Make each decimal an equivalent decimal with two digits after the decimal point in order to easily compare. You then can see that choice (C) is correct: $-0.75 < 0.39 < 0.40 < 0.50$.

9. A

This problem tests your knowledge of complex fractions and fraction division. To divide fractions, multiply the first fraction by the reciprocal, or "flip," of the second. This problem is actually $\frac{2}{3} \div \frac{8}{1}$, or $\frac{2}{3} \times \frac{1}{8}$. Cancel out the common factor of 2 and multiply to get $\frac{1}{12}$.

10. C

This problem tests fraction addition. To add $\frac{1}{3}$ and $\frac{2}{5}$, you need to find a common denominator. Change each fraction to have a common denominator of 15: $\frac{1}{3} = \frac{5}{15}$, $\frac{2}{5} = \frac{6}{15}$. Then, add the numerators and keep the denominator. $\frac{1}{3} + \frac{2}{5} = \frac{5}{15} + \frac{6}{15} = \frac{5+6}{15} = \frac{11}{15}$.

FRACTION AND DECIMAL BASICS

Fractions and decimals are two ways to express numbers. It is important to be comfortable with these notations, and to be adept at performing mathematical operations with both types of numbers. This chapter will cover how to convert from one form of the number to the other, how to order numbers in the two forms, and how to perform the operations of addition, subtraction, multiplication, and division with both types.

Divisibility Rules

In the upcoming lessons on fractions, you will be asked to find factors of a whole number. It is helpful to be familiar with divisibility rules for whole numbers.

REMEMBER THIS!

Divisibility by 2: A number is divisible by 2 if the number is an even number.

Examples: The numbers 24, 50, and 66 are all divisible by 2 because they are all even numbers.

Divisibility by 3: A number is divisible by 3 if the sum of the individual digits in the number is divisible by 3.

Examples: The number 312 is divisible by 3 because $3 + 1 + 2 = 6$, which is divisible by 3. The number 9,021 is divisible by 3 because $9 + 0 + 2 + 1 = 12$, which is divisible by 3.

Divisibility by 4: A number is divisible by 4 if the last two digits, taken as a two-digit number, are divisible by 4.

Examples: The number 736 is divisible by 4 because 36 is divisible by 4. The number 12,716 is divisible by 4 because 16 is divisible by 4.

Divisibility by 5: A number is divisible by 5 if the last digit of the number is a 5 or a 0.

Examples: The numbers 10, 25, 30, and 75 are all divisible by 5 because the last digit of each is a 5 or a 0.

Divisibility by 8: A number is divisible by 8 if the last three digits, taken as a three-digit number, are divisible by 8.

Examples: The number 3,024 is divisible by 8 because 024, or just 24, is divisible by 8. The number 79,128 is divisible by 8 because 128 is divisible by 8; $128 \div 8 = 16$.

Divisibility by 9: A number is divisible by 9 if the sum of the individual digits in the number is divisible by 9.

Examples: The number 9,135 is divisible by 9 because $9 + 1 + 3 + 5 = 18$, which is divisible by 9. The number 414,972 is divisible by 9 because $4 + 1 + 4 + 9 + 7 + 2 = 27$, which is divisible by 9.

Practice 1

Answer true or false for the following statements about divisibility. Answers and explanations are located at the end of the chapter.

1. **T   F**   5,940 is divisible by 3

2. **T   F**   719 is divisible by 9

3. **T   F**   2,528 is divisible by 4

4. **T   F**   1,550 is divisible by 2 and 5

Prime Factors of a Number

In order to understand how to perform operations with fractions, you must be comfortable with the concepts of factors and multiples.

> **MATH SPEAK**
>
> A **factor** of a number x is a whole number that divides into x evenly without remainder.

All the factors of a number can be reduced down into only prime numbers. Every number has a unique set of **prime factors**. You can make a factor tree to find these prime factors. Below are examples of factor trees for the numbers 28 and 80.

$$28 = 2 \cdot 2 \cdot 7 \qquad 80 = 2 \cdot 2 \cdot 2 \cdot 2 \cdot 5$$

As you can see from the example above, there may be more than one way to make a factor tree. No matter how you approach your factor tree, in the end a whole number will break down into the same exact set of prime factors.

Practice 2

Check your understanding of prime factorization by answering the following questions. Answers and explanations are located at the end of the chapter.

5. What is the prime factorization of 120?

6. What is the prime factorization of 40?

The Greatest Common Factor

The greatest common factor (GCF) is the largest common factor of two or more numbers. The GCF is often used to simplify a fraction. Prime factorization is a convenient way to find the GCF of two or more numbers.

After finding the prime factors of the numbers you are working with, place them in a Venn diagram by pairing up all common primes. The Venn diagram that follows illustrates this by placing all common pairs of prime factors of 28 and 80 in the intersection of the circles. The factors without a match are in the outer circles of the Venn diagram.

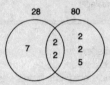

Find the greatest common factor by multiplying only the common factors in the intersection: $2 \times 2 = 4$.

As another example, to find the GCF of 45 and 60, first find the prime factorization of 45 ($3 \times 3 \times 5$) and 60 ($2 \times 2 \times 3 \times 5$). Create the Venn diagram for the factors of 45 and 60:

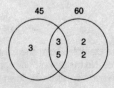

So the GCF of 45 and 60 is $3 \times 5 = 15$.

Practice 3

Use the bank of numbers to answer the following questions. Answers and explanations are located at the end of the chapter.

   6   24   30   16   12   8   4

7. What is the greatest common factor of 24 and 96?

8. What is the greatest common factor of 48 and 100?

The Least Common Multiple

The least common multiple (LCM) is the smallest number that two or more numbers will divide into evenly. The LCM of two or more numbers is used when adding or subtracting fractions. Again, the prime factorization of numbers and the use of a Venn diagram are convenient ways to find the LCM. After creating the diagram, just multiply *all* of the factors appearing in the diagram and you will have the LCM of the two numbers.

Using the preceding diagram for the numbers 28 and 80, the LCM is $(2 \times 2) \times 2 \times 2 \times 5 \times 7 = 560$; the factors in parentheses are the common pairs. Now, look back at the diagram for the numbers 45 and 60. The least common multiple is $2 \times 2 \times 3 \times (3 \times 5) = 180$.

Practice 4

In the following problems, match the pair of numbers on the left with the least common multiple on the right. Answers and explanations are located at the end of the chapter.

 9. 48, 100   A. 60

10. 24, 96   B. 96

11. 12, 60   C. 1,200

12. 27, 90   D. 270

FRACTIONS

> **MATH SPEAK**
>
> A **fraction** is a rational number in the form $\frac{a}{b}$. The variable a is called the numerator, and the variable b is called the denominator. There is one restriction on b: The denominator cannot have the value of 0.

The fractional form of any rational number is the most precise way to represent the number, unless the number is an integer. There are many reasons that numbers are represented as fractions. Some fractions, or ratios, show a comparison of two numbers. This concept will be covered in a future chapter. Fractions can also represent a part-whole relationship. Fractions are usually used to display probabilities, as well.

Equivalent Fractions

A single number can have several equivalent fractional forms.

> **REMEMBER THIS!**
>
> To find an equivalent fraction, just multiply or divide the fraction by the number one, in the special form of $\frac{a}{a}$.

For example, $\frac{5}{35}$ is equivalent to $\frac{1}{7}$ because $\frac{5}{35} \div \frac{5}{5} = \frac{1}{7}$. To find an equivalent fraction to $\frac{3}{4}$, multiply by some fractional form of one: $\frac{3}{4} \times \frac{6}{6} = \frac{18}{24}$. Often, when performing operations on fractions, you will be instructed to simplify a fraction. To simplify a fraction, divide by the greatest common factor of the numerator and the denominator. To simplify $\frac{28}{80}$, divide by $\frac{4}{4}$ because 4 is the GCF. $\frac{28}{80} \div \frac{4}{4} = \frac{7}{20}$.

Practice 5

Use the bank of fractions to choose an equivalent value for each of the following fractions:

$$\frac{7}{5} \quad \frac{4}{5} \quad \frac{4}{7} \quad \frac{20}{30} \quad \frac{45}{60} \quad \frac{60}{90} \quad \frac{14}{25} \quad \frac{18}{25} \quad \frac{8}{25} \quad \frac{15}{63} \quad \frac{15}{42}$$

13. $\dfrac{56}{100}$

14. $\dfrac{3}{4}$

15. $\dfrac{28}{35}$

16. $\dfrac{5}{21}$

Fraction Operations

Multiplication

> **REMEMBER THIS!**
>
> To multiply fractions, simply multiply the numerators and then multiply the denominators.

An example is $\dfrac{7}{9} \times \dfrac{3}{16}$. You can cancel out any common factors found in any of the numerators and denominators to make the multiplication easier:

$$\frac{7}{{}_{3}\cancel{9}} \times \frac{\cancel{3}^{1}}{16} = \frac{7 \times 1}{3 \times 16} = \frac{7}{48}$$

> ## REMEMBER THIS!
> In mathematics, the word "of" most often means to multiply.

To find $\frac{1}{9}$ of $\frac{4}{5}$, just multiply: $\frac{1}{9} \times \frac{4}{5} = \frac{1 \times 4}{9 \times 5} = \frac{4}{45}$.

Practice 6

Try the following fractional multiplication problems. Answers and explanations are located at the end of the chapter.

17. What is $\frac{1}{3}$ of $\frac{4}{7}$?

18. $\frac{5}{8} \times \frac{5}{7} = ?$

19. $\frac{3}{7} \times \frac{7}{11} = ?$

Division

To divide fractions, recall that division is the inverse operation of multiplication. The multiplicative inverse of a fraction is the **reciprocal**, or "flip" of the fraction.

> ## REMEMBER THIS!
> The easiest and most common way to divide fractions is to take the first fraction and multiply it by the **reciprocal** of the second fraction.

This method is effective because you are essentially doing two inverse operations (multiplying and taking the reciprocal) to result in division.

For example, $\frac{5}{8} \div \frac{3}{2}$ is equivalent to $\frac{5}{8} \times \frac{2}{3} = \frac{5 \times 2}{8 \times 3} = \frac{10}{24}$.

Practice 7

Repetition is the key to mastery, so try these fractional division problems. Answers and explanations are located at the end of the chapter.

20. $7 \div \dfrac{2}{3} = ?$

21. $\dfrac{12}{7} \div \dfrac{12}{2} = ?$

22. $\dfrac{8}{9} \div \dfrac{5}{3} = ?$

Addition and Subtraction

To add and subtract fractions, you first need to change the fractions so they have common denominators. The best common denominator to use is the least common multiple of the denominators involved. This is called the **least common denominator**.

REMEMBER THIS!

To add or subtract fractions:

1) Convert the fractions so they have an equivalent **least common denominator**.

2) Add or subtract the numerators and keep the denominator.

3) Simplify the resultant fraction if necessary.

For example, to add $\dfrac{13}{15}$ and $\dfrac{3}{5}$, find the LCM of the denominators.

This is 15. Rewrite the problem with denominators of 15:

$$\frac{13}{15} + \frac{9}{15} = \frac{13 + 9}{15} = \frac{22}{15}.$$

You may also be faced with positive and negative arithmetic. Take $\frac{1}{3} - \frac{19}{20}$. The common denominator in this case is 60. Rewrite the fractions, then subtract the numerators and keep the denominator: $\frac{1}{3} - \frac{19}{20} = \frac{20}{60} - \frac{57}{60}$. In the numerator, change subtraction to addition of the opposite. So you subtract and take the sign of the larger absolute value, resulting in a negative answer. $\frac{20}{60} - \frac{57}{60} = \frac{20 + (-57)}{60} = -\frac{37}{60}$.

Practice 8

Constant practice and repetition are the keys to learning math! Try these two problems for more practice adding and subtracting fractions. Answers and explanations are located at the end of the chapter.

23. $\frac{1}{12} + \frac{5}{8} = ?$

24. $-\frac{6}{7} + \frac{3}{21} = ?$

Mixed Numbers

Mixed numbers are numbers in the form $A\frac{b}{c}$.

REMEMBER THIS!

When performing operations using mixed numbers, it is helpful to first change the mixed number into an improper fraction and then use the correct procedure for fractions as previously described. Change $A\frac{b}{c}$ to an improper fraction by multiplying $c \times A$ and adding it to b. Then put this number over c: $\frac{(A \times c) + b}{c}$.

For example, to multiply, $2\frac{1}{3} \times \frac{1}{4}$, first convert the mixed number to an improper fraction: $2\frac{1}{3} = \frac{(2 \times 3 + 1)}{3} = \frac{7}{3}$. Then, multiply the numerators and multiply the denominators: $\frac{7}{3} \times \frac{1}{4} = \frac{7 \times 1}{3 \times 4} = \frac{7}{12}$.

Similarly, to add $4\frac{3}{8} + -2\frac{7}{8}$, first change the mixed numbers to improper fractions. The fractions have the same denominator, so add the numerators and keep the denominator of 8: $4\frac{3}{8} + -2\frac{7}{8} = \frac{35}{8} + -\frac{23}{8} = \frac{35 - 23}{8} = \frac{12}{8}$. Simplify by canceling the common factor of 4 to get $\frac{3}{2}$.

Practice 9

Practice operations with mixed numbers in the following questions. Answers and explanations are located at the end of the chapter.

25. $11\frac{2}{3} \div 3 = ?$

26. $3\frac{1}{3} - 12\frac{5}{6} = ?$

27. Kafi wants to make bread stuffing for 15 people, but the recipe she has serves only 6. How many cups of bread does she need to serve 15 if the recipe calls for $4\frac{1}{2}$ cups of bread?

Complex Fractions

> **MATH SPEAK**
>
> **Complex fractions** are fractions whose numerator or denominator is also a fraction.

To simplify these fractions, remember that the fraction bar means to divide. Rewrite the fraction as a division problem, and follow the procedure for dividing fractions. To simplify $\dfrac{\frac{8}{15}}{4}$, rewrite it as

division: $\dfrac{8}{15} \div 4 = \dfrac{8}{15} \times \dfrac{1}{4} = \dfrac{8}{60} = \dfrac{2}{15}$.

Practice 10

Try simplifying these complex fractions. Answers and explanations are located at the end of the chapter.

28. $\dfrac{\frac{9}{5}}{\frac{3}{10}} = ?$

29. $\dfrac{\frac{2}{3}}{\frac{1}{2}} = ?$

DECIMALS

Decimal form is a convenient way to express a number, because it is based on the powers of 10. This makes comparisons like addition, subtraction, multiplication, and division easier to perform.

MATH SPEAK

Look at the decimal form "*abc.def*" to understand place value.

The letter *a* is the digit in the hundreds place, the letter *b* is in the tens place, *c* is in the ones place; *d* is in the tenths $\left(\dfrac{1}{10}\right)$ place, *e* is in the hundredths place $\left(\dfrac{1}{100}\right)$, and *f* is in the thousandths place $\left(\dfrac{1}{1000}\right)$.

As an example, 5.092 is equivalent to 5 and 92 thousandths because the rightmost digit is three places after the decimal point.

Practice 11

For the set of following statements, fill in the blank with the correct decimal place value word. Answers are located at the end of the chapter.

30. In the decimal number 3.507, the 5 is in the _____ place.

31. In the number 17.624 the 2 is in the _____ place.

32. The decimal 9.003 represents 9 and three- _____.

Equivalent Decimals

The number 3.5 is equivalent to any other decimal form of that number with trailing zeroes; 3.5, 3.500, and 3.50000 are all equivalent decimal numbers. Each represents 3 and five-tenths. When adding, subtracting, or comparing decimals it is often convenient to add trailing zeroes, as the lessons following will demonstrate.

Conversions

> ### REMEMBER THIS!
>
> To convert a decimal to a fraction or mixed number, just use the place value and then simplify if necessary. The decimal 16.4 is equal to $16\frac{4}{10}$, or $16\frac{2}{5}$ simplified. To change a fraction to a decimal, recognize that a fraction bar means to divide. So to change $\frac{6}{15}$ to a decimal, divide 6 by 15 to get 0.4.

To change a mixed number to a decimal, separate and keep the whole number part; this is the number to the left of the decimal point. Then divide the fractional part as described above. The mixed number $12\frac{3}{4}$ is 12.75 because 3 divided by 4 is 0.75.

Practice 12

Match each fraction on the left with its decimal equivalent on the right. Answers are located at the end of the chapter.

33. $3\frac{1}{5}$ A. 3.4

34. $\frac{3}{20}$ B. 3.20

35. $3\frac{3}{25}$ C. 0.15

36. $3\frac{2}{5}$ D. 3.12

Ordering

> ### REMEMBER THIS!
> To order or compare fractions, first convert the fractions to decimals by dividing, and then order the decimal equivalents.

The decimal form of a number makes comparisons easy. Just rewrite each decimal to have the same number of places to the right of the decimal point and then compare the numbers. For example, to order the decimals {1.2, 1.07, 1.019} from least to greatest, rewrite the numbers as 1.200, 1.070, and 1.019. Since the whole number parts are the same, just order the decimal portion. With the trailing zeroes it is evident that 200 > 70 > 19, so the correct order from least to greatest is 1.019, 1.07, 1.2.

Practice 13

For each list, tell whether the fractions and decimals are in increasing order, decreasing order, or no order. Answers are located at the end of the chapter.

37. $\dfrac{3}{7}, \dfrac{2}{3}, \dfrac{3}{4}, \dfrac{5}{6}$

38. 7.5, 7.47, 7.08, 7.009

39. $-\dfrac{3}{4}, \dfrac{1}{8}, \dfrac{7}{16}, \dfrac{1}{2}$

40. −0.16, −0.42, 1.5, 1.83

Decimal Operations

Addition and Subtraction

> **REMEMBER THIS!**
>
> To add or subtract decimal numbers, remember that it is
> imperative to LINE UP the decimal points! This is the FIRST step
> before performing addition or subtraction. Add trailing zeroes if
> necessary to avoid careless mistakes.

For example, to add 2.509 to 234.6, first line up the decimal points as
shown below. Two trailing zeroes are added to the end of 234.6 and
then addition is performed as shown.

$$
\begin{array}{r}
234.600 \\
+2.509 \\
\hline
237.109
\end{array}
$$

> **FLASHBACK**
>
> The positive and negative arithmetic rules that you reviewed in
> chapter 2 with integers also apply to decimal numbers.

To subtract 5.908 from 3.7 (that is 3.7 − 5.908), recall that subtraction
is the same as adding the opposite. Change the problem to read
3.7 + −5.908. To add two numbers with different signs, subtract the
number with the smaller absolute value from the number with the
larger absolute value. The sign of the answer will be the sign of the
number that has the larger absolute value. The figure on the following
page shows this subtraction. The final answer is negative because
−5.908 is negative.

$$
\begin{array}{r}
5.908 \\
-\ 3.700 \\
\hline
2.208
\end{array}
$$

The answer is -2.208
because 5.908 was negative.

Practice 14

Use the number bank below to find the decimal sums and differences. Answers and explanations are located at the end of the chapter.

9.27	113.15	−2.16	17.96	10172
11.315	22.9	1.3	7.16	9.24

41. −1.04 + 8.2

42. 12.7 + 100.45

43. 24 − 1.1

44. 7.9 − 10.06

Multiplication

> **REMEMBER THIS!**
>
> To multiply decimal numbers, follow the steps below:
>
> 1. Multiply the numbers without regard to the decimal point and obtain a whole number product.
>
> 2. Count the number of digits that are to the right of the decimal point in BOTH factors.
>
> 3. Alter the whole number product to have the same number of digits to the right of the decimal point, as counted in step 2.

For example, to multiply 12.9 times 0.07, step 1 instructs you to multiply 129 times 7, which equals 903. In step 2, count the number of digits to the right of both decimal points; there is one digit in 12.9 and there are two digits in 0.07: $1 + 2 = 3$. In step 3, take the whole number product, 903, and alter it so that it has three digits to the right of the decimal point, making the answer 0.903.

FLASHBACK

Multiplication with negative numbers will follow the same rule described in chapter 2.

Practice 15

45. The product resulting from 3.098 times 1.02 will have _____ digits to the right of the decimal point.

46. The product resulting from 1,254 times 234.003 will have _____ digits to the right of the decimal point.

47. $1.002 \times 7.1 = ?$

48. $922 \times 0.06 = ?$

Division

REMEMBER THIS!

To divide decimal numbers, follow the steps below:

1. Set up the long division problem.

2. Count how many digits are to the right of the decimal point in the divisor (the number you are dividing with).

3. Move the decimal point in the dividend (the number you are dividing into) the amount from step 2.

4. Raise the newly placed decimal point up to the quotient.

5. Divide as usual as if there were no decimal points.

For example, to divide 0.39 by 2.6, look at the divisor. 2.6 has one digit to the right of the decimal point. Therefore you move the decimal point one place to the right in the dividend. Move this new point position up to the quotient. The long division is shown in the figure below:

$$
\begin{array}{r}
.15 \\
2.6\,\overline{)0.390} \\
\underline{26} \\
130 \\
\underline{130} \\
0
\end{array}
$$

$0.39 \div 2.6 = 0.15$

Practice Set 16

49. $960 \div 0.08 = ?$

50. $5.67 \div 1.35 = ?$

SUMMARY

To review, here are five things about fractions you should take from this chapter:

- To simplify a fraction, divide the numerator and the denominator by the greatest common factor.
- To multiply fractions, multiply the numerator, multiply the denominator, and then simplify.
- To divide fractions, change the division operation to multiplication, change the divisor (second) fraction to its reciprocal (or "flip"), and multiply.
- To add or subtract fractions, change each fraction to have a common denominator. Add or subtract the numerators and keep the denominator. Simplify if needed.
- To change a fraction into a decimal, divide the numerator by the denominator.

Here are five things about decimals you should take from this chapter:

- Decimals are a convenient form of numeration that can be used to compare numbers.

- To change a decimal to a fraction, write it as such based on the place value of the digits, and then simplify the fractional part if necessary.
- To add or subtract decimals, line up the decimal points in the addends and in the sum or difference.
- To multiply decimals, multiply as whole numbers. Count the number of digits that appear to the right of the decimal point in the two factors. Alter the product so that it has this number of digits to the right of the decimal point.
- To divide decimals, move the decimal point in the dividend and quotient the number of places that you need to move to make the divisor a whole number. Then, divide as usual.

Practice Answers and Explanations

1.   **True**

$5 + 9 + 4 + 0 = 18$, which is divisible by 3.

2.   **False**

$7 + 1 + 9 = 17$, which is not divisible by 9.

3.   **True**

28 is divisible by 4.

4.   **True**

1,550 is even, and ends in a 0.

5. $2 \times 2 \times 2 \times 3 \times 5$

The prime factorization of 120 is shown in the tree.

6. $2 \times 2 \times 2 \times 5$

The prime factorization of 40 is shown in the tree.

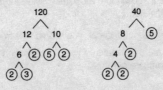

$$120 = 2 \cdot 2 \cdot 2 \cdot 3 \cdot 5 \qquad 40 = 2 \cdot 2 \cdot 2 \cdot 5$$

7. 24

The prime factorization of 24 is $2 \times 2 \times 2 \times 3$ and of 96 is $2 \times 2 \times 2 \times 2 \times 2 \times 3$. The factors in common are $2 \times 2 \times 2 \times 3 = 24$.

8. 4

The prime factorization of 48 is $2 \times 2 \times 2 \times 2 \times 3$ and of 100 is $2 \times 2 \times 5 \times 5$. The factors in common are $2 \times 2 = 4$.

9. C

The Venn diagram is shown, with common factors in the intersection of the circles. Multiply all factors shown in the diagram to get $2 \times 2 \times 2 \times 2 \times 3 \times 5 \times 5 = 1,200$.

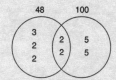

$$48 = 2 \cdot 2 \cdot 2 \cdot 2 \cdot 3 \qquad 100 = 2 \cdot 2 \cdot 5 \cdot 5$$

10. B

Look at the Venn diagram below. All of the factors of 24 are in the common intersection area because they are also factors of 96. The LCM is 96.

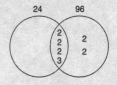

$$24 = 2 \cdot 2 \cdot 2 \cdot 3 \qquad 96 = 2 \cdot 2 \cdot 2 \cdot 2 \cdot 2 \cdot 3$$

11. A

Since $12 \times 5 = 60$, 60 is a multiple of 12. The least common multiple is 60.

12. D

The prime factorization of 27 is $3 \times 3 \times 3$ and of 90 is $3 \times 3 \times 5 \times 2$. So the LCM is the common pairs (3×3) multiplied by the leftovers $(3 \times 5 \times 2)$. $3 \times 3 \times 3 \times 5 \times 2 = 270$.

13. $\dfrac{14}{25}$

$$\frac{56}{100} \div \frac{4}{4} = \frac{14}{25}$$

14. $\dfrac{45}{60}$

$$\frac{45}{60} \div \frac{15}{15} = \frac{3}{4}$$

15. $\dfrac{4}{5}$

$$\frac{28}{35} \div \frac{7}{7} = \frac{4}{5}$$

16. $\dfrac{15}{63}$

$$\frac{5}{21} \times \frac{3}{3} = \frac{15}{63}$$

17. $\dfrac{4}{21}$

The key word "of" means to multiply. $\dfrac{1}{3} \times \dfrac{4}{7} = \dfrac{1 \times 4}{3 \times 7} = \dfrac{4}{21}$.

18. $\dfrac{25}{56}$

$$\frac{5}{8} \times \frac{5}{7} = \frac{5 \times 5}{8 \times 7} = \frac{25}{56}.$$

19. $\dfrac{3}{11}$

The factor of 7 can be canceled in the denominator of the first factor and the numerator of the second factor.

20. $\dfrac{21}{2}$

Change the problem to multiplication and take the reciprocal of the divisor: $7 \div \dfrac{2}{3} = \dfrac{7}{1} \times \dfrac{3}{2} = \dfrac{7 \times 3}{1 \times 2} = \dfrac{21}{2}$.

21. $\dfrac{2}{7}$

Change the problem to multiplication and take the reciprocal of the divisor. The factor of 12 can be canceled, then multiply: $\dfrac{12}{7} \times \dfrac{2}{12} = \dfrac{1 \times 2}{7 \times 1} = \dfrac{2}{7}$.

22. $\dfrac{8}{15}$

Change the problem to multiplication and take the reciprocal of the divisor. $\dfrac{8}{9} \div \dfrac{5}{3} = \dfrac{8}{9} \times \dfrac{3}{5}$. The factor of 3 can be canceled, changing the 3 in the numerator to a 1 and the 9 in the denominator to a 3: $\dfrac{8}{9} \times \dfrac{3}{5} = \dfrac{8 \times 1}{3 \times 5} = \dfrac{8}{15}$.

23. $\dfrac{17}{24}$

Alter both fractions so they have a denominator of 24, the LCM of the denominators. $\dfrac{1}{12} + \dfrac{5}{8} = \dfrac{2}{24} + \dfrac{15}{24} = \dfrac{17}{24}$.

24. $-\dfrac{15}{21}$

Change $\dfrac{6}{7}$ to have a denominator of 21, then add the fractions with different signs. This means to subtract and take the sign of the number with the larger absolute value: $-\dfrac{6}{7} + \dfrac{3}{21} = -\dfrac{18}{21} + \dfrac{3}{21} = \dfrac{-18 + 3}{21} = \dfrac{15}{21}$.

25. $\frac{35}{9}$

First change the mixed number to a fraction; then alter the problem to

multiply by the reciprocal of 3: $\frac{35}{3} \times \frac{1}{3} = \frac{35 \times 1}{3 \times 3} = \frac{35}{9}$.

26. $-\frac{57}{6}$

First change the mixed numbers to fractions. $3\frac{1}{3} = \frac{(9+1)}{3} = \frac{10}{3}$.

$12\frac{5}{6} = \frac{(72+5)}{6} = \frac{77}{6}$. Then alter the fractions so that each has 6 as

the common denominator. Subtract, and take the sign of the larger

absolute value: $\cdot\frac{10}{3} - \frac{77}{6} = \frac{20}{6} - \frac{77}{6} = \frac{20-77}{6} = -\frac{57}{6}$

27.   11.25 cups

Here is an example of a time when a fraction is also a ratio. Kafi wants

to serve 15, but her recipe only serves 6. Put $\frac{15}{6}$ to show this

comparison. When you divide 15 by 6, you get $2\frac{1}{2}$. Therefore, anything

fit to serve 6 people, when multiplied by $2\frac{1}{2}$, will then serve 15

people. So all ingredient amounts must be multiplied by $2\frac{1}{2}$. To find

the amount of bread needed, muliply: $4\frac{1}{2} \times 2\frac{1}{2} = \frac{9}{2} \times \frac{5}{2} = \frac{45}{4}$. This

is equivalent to $11\frac{1}{4}$, or 11.25 cups of bread.

28.   6

Write the problem as a division problem: $\frac{9}{5} \div \frac{3}{10}$. Change the

problem to multiply by the reciprocal. Cancel out the common factors

of 3 and 5. Finally, multiply across to get the simplified answer:

$\frac{9}{5} \times \frac{10}{3} = \frac{3}{1} \times \frac{2}{1} = 6$.

29. $\frac{4}{3}$

Write the problem as a division problem: $\frac{2}{3} \div \frac{1}{2}$. Change the problem to multiply by the reciprocal. Multiply across to get the simplified answer: $\frac{2}{3} \times \frac{2}{1} = \frac{4}{3}$.

30. tenths

31. hundredths

32. thousandths

33. B

34. C

35. D

36. A

37. increasing

38. decreasing

39. increasing

40. no order

41. 7.16

When you add two numbers with different signs, subtract the numbers and take the sign of the larger absolute value. Before subtracting, add a trailing zero to 8.2 so that the decimal point will be lined up correctly. $8.20 - 1.04 = 7.16$.

42. 113.15

Add a trailing 0 onto 12.7 to get 12.70 so that the numbers will be aligned with the decimal points lined up. $12.70 + 100.45 = 113.15$.

43. 22.9

Add a 0 in the tenths place to the whole number 24, to get 24.0. Now subtract, making sure that the decimal point in lined up. $24.0 - 1.1 = 22.9$.

44. −2.16

When you subtract, change the problem to addition and change the sign of the second addend. The problem becomes 7.9 + -10.06. When you add two numbers with different signs, subtract the smaller from the larger, and keep the sign of the number with the larger absolute value. 10.06 − 7.90 = 2.16, and this result will be negative since 10.06 is negative.

45. five

3.098 has 3 digits to the right of the decimal point; 1.02 has 2 digits. 3 + 2 = 5.

46. three

1,254 has 0 digits to the right of the decimal point; 234.003 has 3. 0 + 3 = 3.

47. 7.1142

The product will have four digits to the right of the decimal point because the first factor has three digits and the second factor has 1 digit to the right of the decimal point: 3 + 1 = 4.

48. 55.32

The product will have 2 digits to the right of the decimal point because the first factor has 0 digits and the second factor has 2 digits to the right of the decimal point.

49. 12,000

There are two digits to the right of the decimal point in the divisor (0.08), so the decimal point is moved over 2 places to the right in the dividend (960). Two zeroes are tacked onto the end of the dividend to make the number 96,000 and the decimal point's new position moves up. 96,000 ÷ 8 = 12,000, because 96 ÷ 8 = 12.

50. 4.2

Both the divisor and the dividend have two places to the right of the decimal point, so move the decimal point in both. The division is now $567 \div 135$. The figure below shows the long division:

$$
\begin{array}{r}
4.2 \\
1.35{\overline{)5.67\,0}} \\
\underline{5.40} \\
27\,0 \\
\underline{27\,0} \\
0
\end{array}
\qquad 5.67 \div 1.35 = 4.2
$$

CHAPTER 3 TEST

Now that you have studied factors, multiples, fractions, and decimals, try the following set of questions. Use the chapter material, including the practice questions throughout, to assist you in solving these problems. The answer explanations that follow will provide additional help.

1. $6.002 - 3.7$
    (A) 5.965   (B) 2.302   (C) 3.005
    (D) 0.2302   (E) 0.5965

2. What is the greatest common factor of 24 and 90?
    (A) 6   (B) 2   (C) 3   (D) 180   (E) 360

3. What is the prime factorization of 120?
    (A) $2 \times 3 \times 5$ (B) $2 \times 5 \times 12$ (C) $3 \times 4 \times 5$
    (D) $2^3 \times 3 \times 5$ (E) $2^2 \times 3 \times 5$

4. Simplify: 2.8×4.002
    (A) 112.056   (B) 6.802   (C) 11.2056
    (D) 1.202   (E) 1.429

5. Which order below shows the decimals from smallest to largest?

   (A) −3.05, −4.2, 1.5, 2.04, 2.009

   (B) −6.09, −2.6, 1.5, 2.009, 2.04

   (C) −7.12, −8.02, 0, 5.2, 10.4

   (D) 2.009, 2.04, 1.5, −4.2, −3.05

   (E) 2.04, 2.009, 1.5, −4.2, −3.05

6. What is $\frac{3}{4}$ of $\frac{2}{3}$?

   (A) $\frac{5}{7}$ (B) $\frac{17}{12}$ (C) $\frac{9}{8}$ (D) $\frac{1}{2}$ (E) $\frac{8}{9}$

7. Which of the following shows the fractions in increasing order?

   (A) $-\frac{1}{4}, -\frac{2}{5}, \frac{2}{3}, \frac{1}{2}$

   (B) $-\frac{2}{5}, -\frac{1}{4}, \frac{1}{2}, \frac{2}{3}$

   (C) $-\frac{1}{4}, -\frac{2}{5}, \frac{1}{2}, \frac{2}{3}$

   (D) $\frac{2}{3}, \frac{1}{2}, -\frac{1}{4}, -\frac{2}{5}$

   (E) $\frac{2}{3}, \frac{1}{2}, -\frac{2}{5}, -\frac{1}{4}$

8. What is the least common multiple of 54 and 32?

   (A) 2   (B) 1   (C) 108   (D) 864   (E) 8

9. The number 11,550 is divisible by:

   (A) 5 only   (B) 3 only   (C) 2 only

   (D) all of the above   (E) none of the above

10. Which fraction is equivalent to $\frac{21}{56}$?

   (A) $\frac{3}{7}$ (B) $\frac{7}{8}$ (C) $\frac{3}{8}$ (D) $\frac{42}{106}$ (E) $\frac{11}{28}$

11. $-\frac{5}{12} \times \frac{6}{24} =$

   (A) $-\frac{5}{48}$ (B) $\frac{5}{48}$ (C) $\frac{10}{6}$

   (D) $\frac{1}{2}$ (E) $-\frac{10}{6}$

12. $3\frac{5}{8} - 4\frac{1}{4} =$

   (A) $\frac{14}{8}$ (B) $\frac{112}{25}$ (C) $\frac{8}{14}$ (D) $-\frac{5}{8}$ (E) $-\frac{14}{8}$

13. $4.212 \div 2.34 =$
   (A) 18   (B) 1.8   (C) 1.08   (D) 0.18   (E) 9.85608

14. $\frac{16}{25} \div -\frac{4}{5} =$

   (A) $\frac{4}{5}$ (B) $-\frac{4}{5}$ (C) $\frac{12}{20}$ (D) $-\frac{12}{20}$ (E) $-\frac{5}{4}$

15. $17.6 \times 0.005 =$
   (A) 0.88   (B) 8.8   (C) 0.0088   (D) 0.088   (E) 3,520

16. Which of the following is NOT true?

   (A) $\frac{1}{4} < \frac{1}{3}$ (B) $-\frac{2}{3} > -\frac{5}{6}$ (C) $\frac{12}{15} = \frac{60}{75}$

   (D) $\frac{3}{4} < \frac{4}{6}$ (E) $\frac{5}{6} \geq \frac{15}{18}$

17. Andrea needs $3\frac{3}{4}$ cups of flour to make 24

    cupcakes. How much flour is needed to make 8 cupcakes?

    (A) $11\frac{1}{4}$ (B) $1\frac{1}{4}$ (C) 30 (D) $\frac{15}{32}$ (E) $1\frac{7}{8}$

18. In a bag of trail mix, $\frac{3}{8}$ of the mix is peanuts.

    How much of a 15 pound trail mix is not peanuts?

    (A) 5.625 pounds (B) 40 pounds (C) 12 pounds
    (D) 7 pounds (E) 9.375 pounds

19. Adult admission to the movie theater is $6.75. The price of a
    ticket for a child under the age of 12 is $3.50. What is the total
    price for 3 adults and 1 child to attend the movie?

    (A) $ 17.25 (B) $ 27.00 (C) $ 23.75
    (D) $ 20.25 (E) $ 206.00

20. Gloria's estate is to be divided evenly among her 3 children.
    One son, Miguel, is deceased; that portion is then to be
    divided among Miguel's 5 children. What fraction of Gloria's
    estate would each of Miguel's children receive?

    (A) $\frac{1}{3}$ (B) $\frac{5}{3}$ (C) $\frac{5}{15}$ (D) $\frac{1}{15}$ (E) $\frac{1}{5}$

21. At the mall one weekend, $\frac{2}{9}$ of the shoppers used a personal

    check to purchase items and $\frac{2}{5}$ of the shoppers paid with

    cash. The rest of the shoppers used credit cards. What

    fraction of the shoppers used credit cards?

    (A) $\frac{28}{45}$ (B) $\frac{41}{45}$ (C) $\frac{5}{9}$ (D) $\frac{5}{7}$ (E) $\frac{17}{45}$

Answers and Explanations

1.   B

Add two zeroes onto the end of the decimal 3.7 so that the two numbers have the same amount of digits to the right of the decimal point. Then, line up your decimal points and subtract to get: $6.002 - 3.700 = 2.302$.

2.   A

The prime factors of 24 are $2^3 \times 3$. The prime factors of 90 are: $2 \times 3^2 \times 5$. The common factors are $2 \times 3 = 6$, the greatest common factor. Choice (B) is a factor of both numbers, but is not the greatest common factor. Choice (C) is also a factor of both numbers, but again, is not the greatest. Choice (D) is a multiple of 90. Choice (E) is the least common multiple of 24 and 90.

3.   D

The number 120 can be broken down into prime factors as follows: first, $120 = 12 \times 10$; $10 = 2 \times 5$; $12 = 6 \times 2$; $6 = 2 \times 3$. This is a total of three factors of 2, one factor of 3 and one factor of 5. Choice (A) is the prime factorization of 30: $2 \times 3 \times 5 = 30$. Choice (B) is a factorization of 120, but 12 is not prime. Choice (C) is a factorization of 60, and 4 is not prime. Choice (E) is the prime factorization of 60.

4.   C

Multiply 4,002 by 28 to get 112,056. Because there are a total of four digits to the right of the decimal points in the factors, place the decimal point in the answer so that there are four places to the right of the decimal point: 11.2056.

5.   B

Choice B shows the decimals arranged from smallest to largest. Even though the absolute value of −6.09 is greater than the absolute value of −2.6, $-6.09 < -2.6$ because they are negative numbers.

6.   D

Remember, the key word "of" means to multiply. When you multiply these fractions, you can cancel a 3 from a numerator and a denominator, as well as canceling a 2 out of the numerator and a denominator. After canceling and multiplying across, the answer is $\frac{1}{2}$.

7.   B

Increasing order shows the fractions ordered from smallest to largest. To make comparison easier, convert all the fractions to decimals. In choice (B), the numbers are: {-0.40, -0.25, 0.50, 0.67}. Because each number has the same amount of digits to the right of the decimal point, it is easy to see that these numbers are ordered correctly.

8.   D

The only common factor of 54 and 32 is 2. Make a Venn diagram as shown below; put the common factor of 2 in the intersection. Put all leftover factors in the outer circles and multiply all numbers in the diagram. The least common multiple is $3^3 \times 2 \times 2^4$. Choice (A) is the greatest common factor of 54 and 32. Choice (B) is a factor of every number. Choice (C) is a multiple of 54, but not a multiple of 32. Choice (E) is a factor of 32, but not of 54.

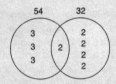

$$54 = 2 \cdot 3 \cdot 3 \cdot 3 \quad 32 = 2 \cdot 2 \cdot 2 \cdot 2 \cdot 2$$

9.   D

The number 11,550 is divisible by 2, 3, and 5. It is an even number, hence divisible by 2. The last digit is a 0, hence it is divisible by 5. The sum of the digits is $1 + 1 + 5 + 5 + 0 = 12$, which is divisible by 3. Choices (A), (B), and (C) are all true.

10.   C

Simplify the fraction by dividing each part by the GCF of 7:

$$\frac{21}{56} \div \frac{7}{7} = \frac{3}{8}.$$

11. A

Cancel the common factor of 6 from a numerator and a denominator,

then multiply to get: $-\dfrac{5}{2} \times \dfrac{1}{24} = -\dfrac{5 \times 1}{2 \times 24} = -\dfrac{5}{48}$. A negative number

times a positive number is a negative number, so the answer is
negative.

12. D

First, change the mixed numbers to improper fractions:

$3\dfrac{5}{8} = \dfrac{(24+5)}{8} = \dfrac{29}{8}. \ 4\dfrac{1}{4} = \dfrac{(16+1)}{4} = \dfrac{17}{4}$. Then, change the second

fraction to have the common denominator of 8: $\dfrac{17}{4} = \dfrac{34}{8}$. Subtract

the numerators and keep the denominator. $\dfrac{29}{8} - \dfrac{34}{8} = \dfrac{29-34}{8} = -\dfrac{5}{8}$.

The answer is negative because $\dfrac{34}{8}$ has the greater absolute value.

13. B

The divisor has two digits to the right of the decimal point, so move the
decimal point in the dividend, 4.212, over two places right. Place this
decimal point up to the quotient. Now, perform long division as shown
below:

$$
\begin{array}{r}
1.8 \\
2.34\overline{)4.21\,2} \\
2\,34 \\
\overline{1\,87\,2} \\
1\,87\,2 \\
\overline{0}
\end{array}
$$

$4.212 \div 2.34 = 1.8$

14. B

Change the problem to multiplication by the reciprocal, that is:

$\dfrac{16}{25} \times -\dfrac{5}{4}$. You can cancel out the common factor of 4 and 5 from a

numerator and a denominator to get $-\dfrac{4}{5}$. A positive divided by a

negative number is a negative number.

15. D

First, multiply as if there were no decimal points: $176 \times 5 = 880$. Because there are a total of four digits to the right of the decimal point in the factors, there must be four digits to the right of the decimal point in the product. This will require you to append a 0 prior to the 880 to get 0.0880.

16. D

Be careful! This problem is asking which of the statements is NOT true. You can see that $\frac{3}{4}$ is greater than $\frac{4}{6}$ by converting the fractions to decimal equivalents: $\frac{3}{4} = 0.75$ and $\frac{4}{6} = 0.67$. $0.75 > 0.67$. Choice (A) is true. $\frac{1}{4} = 0.25$, which IS less than $\frac{1}{3} = 0.\overline{3}$. Choice (B) is true. $-\frac{2}{3} = -0.\overline{6}$, which IS greater than $-\frac{5}{6} = -0.8\overline{3}$, because they are negative numbers. Choice (C) is true. $\frac{12}{15} \times \frac{5}{5} = \frac{60}{75}$. Choice (E) is true. The inequality is read as "greater than or equal to" and $\frac{5}{6} = \frac{15}{18}$, because $\frac{15}{18} \div \frac{3}{3} = \frac{5}{6}$.

17. B

First, set up the ratio of the values given: $\frac{24}{8}$. This problem is essentially asking you to take $3\frac{3}{4}$, or $\frac{15}{4}$, and divide by 3, because $24 \div 8 = 3$. Change division to multiplication by the reciprocal: $\frac{15}{4} \div \frac{3}{1} = \frac{15}{4} \times \frac{1}{3}$. Now, simplify by dividing out the common factor of 3 to get $\frac{5}{4} = 1\frac{1}{4}$ cups of flour.

18. E

The key word "of" means to multiply. Multiply $\frac{3}{8}$ by 15: $\frac{3}{8} \times \frac{15}{1} = \frac{45}{8}$.

$45 \div 8 = 5.625$ pounds of mix. Be sure to read the question carefully.

This the amount of mix that IS peanuts. The amount of mix that is NOT

peanuts is $15.000 - 5.625 = 9.375$.

19. C

There are three adults and one child buying tickets. The total price is
$(3 \times 6.75) + 3.50$. $6.75 \times 3 = 20.25$. $20.25 + 3.50 = \$23.75$.

20. D

Each of Gloria's children gets $\frac{1}{3}$ of the estate. Miguel's $\frac{1}{3}$ portion will

be divided by 5. When you divide, you can multiply by the reciprocal of

the divisor to get $\frac{1}{3} \times \frac{1}{5} = \frac{1}{15}$.

21. E

First, add $\frac{2}{9} + \frac{2}{5}$. Find the common denominator of 45. Add the

numerators and keep the denominators, to get $\frac{10}{45} + \frac{18}{45} = \frac{28}{45}$. Read

the question carefully. This is the fraction of people who do NOT use

credit. The fraction of people who DO use credit is $\frac{45}{45} - \frac{28}{45} = \frac{17}{45}$.

CHAPTER 4

Ratio and Proportion

BUILDING BLOCK QUIZ

This Building Block Quiz for chapter 4 includes some questions to help you review the material from chapter 3, Fractions and Decimals. The quiz will also serve as a warm-up for the concepts of ratio and proportion that will be presented in this chapter. Detailed answer explanations are provided to help you find your errors and clarify concepts.

1. Which of the following is equivalent to $\frac{8}{12}$?

   (A) $\frac{2}{3}$ (B) $\frac{4}{5}$ (C) $\frac{2}{6}$ (D) $\frac{7}{11}$ (E) $\frac{14}{20}$

2. Multiply: $2\frac{1}{3} \times 3\frac{1}{7}$

   (A) $7\frac{1}{3}$ (B) $6\frac{1}{21}$ (C) $5\frac{2}{21}$ (D) $6\frac{1}{5}$ (E) $3\frac{2}{21}$

3. The decimal 0.625 can also be expressed as which of the following?

   (A) $\frac{7}{8}$ (B) $\frac{5}{6}$ (C) $\frac{4}{5}$ (D) $\frac{5}{8}$ (E) $\frac{3}{5}$

4. The ratio 3 to 5 can also be expressed as the ratio 15 to _____?

   (A) 9 (B) 12 (C) 17 (D) 25 (E) 45

5. The ratio of boys to girls in the Drama Club is 4 : 7. What is the ratio of girls to the total number of students in the Drama Club?

(A) $\frac{4}{7}$ (B) $\frac{7}{4}$ (C) $\frac{7}{11}$ (D) $\frac{4}{11}$ (E) $\frac{11}{4}$

6. Solve for x: $\frac{8}{9} = \frac{x}{108}$.

(A) 12   (B) 72   (C) 81   (D) 96   (E) 144

7. A copy machine copies 68 sheets per minute. At this rate, how many copies can it make in 8.5 minutes?

(A) 8   (B) 68   (C) 544   (D) 572   (E) 578

8. A scale model of an airplane has a scale of 1:16. If the model is 38 inches long, what is the length of the actual airplane, in feet?

(A) 16 ft 10 in   (B) 38 ft   (C) 50 ft 6 in

(D) 50 ft 8 in   (E) 608 ft

9. The corresponding sides of two similar triangles measure 10 and 15 meters, respectively. What is the perimeter of the smaller triangle if the perimeter of the larger triangle is 34.5 meters?

(A) 19.5 m   (B) 23 m   (C) 25 m   (D) 51 m   (E) 52 m

10. If n dinner rolls cost the same as m bagels, and bagels cost 56 cents each, then how many cents does each dinner roll cost?

(A) $\frac{56m}{n}$ (B) $\frac{56n}{m}$ (C) $56mn$

(D) $\frac{1}{56mn}$ (E) $\frac{n}{56m}$

Answers and Explanations

1.   A

This problem tests the concept of equivalent fractions from chapter 3. One approach to this question is to simplify the fraction $\frac{8}{12}$ to $\frac{2}{3}$ by dividing both the numerator and denominator by the greatest common factor of 4.

2.   A

First, change each of the mixed numbers to an improper fraction. The problem becomes $\frac{7}{3} \times \frac{22}{7}$. Cancel the common factors of 7 in both the numerator and denominator. Multiply across to get $\frac{22}{3}$, which is equal to $7\frac{1}{3}$ as a mixed number.

3.   D

This question also tests a concept from chapter 3—changing a decimal to a fraction. Because it is a number that extends three places to the right of the decimal point, the number is read as "six-hundred-twenty-five thousandths." The fraction equivalent to this is $\frac{625}{1,000}$. Divide both numerator and denominator by the GCF of 125 to simplify the fraction to $\frac{5}{8}$.

4.   D

Questions 4 and 5 each test the concept of ratios. To compare the ratio 3 to 5 with another ratio, write the ratios in fraction form and set them equal to one another. $\frac{3}{5} = \frac{15}{?}$. Since $3 \times 5 = 15$ in the numerators, the values in the second ratio are 5 times larger than the first ratio. To find the solution, multiply 5×5 to get 25 in the denominator.

5. C

Because the label boys is mentioned first in the sentence, the first number listed corresponds with the number of boys in the Drama Club. Thus, 4 corresponds with the number of boys and 7 corresponds with the number of girls. The total number of students would then be represented by the value $4 + 7 = 11$. Since the question asks for the ratio of girls to the total number of students, the final ratio would then be 7 to 11, which can also be written as $\frac{7}{11}$.

6. D

Question 6 tests your ability to solve proportions. To solve this proportion, cross-multiply the numerator of the first fraction and the denominator of the second fraction and set it equal to the product of the denominator of the first fraction with the numerator of the second fraction. Another way to say this is to multiply the *means* by the *extremes*. The equation becomes $864 = 9x$. Divide each side by 9 to get $x = 96$.

7. E

This problem tests the concept of rate. Since the rate known in the problem is 68 sheets per minute, it is a unit rate and is the same as $\frac{68 \text{ sheets}}{1 \text{ minute}}$. To find the number of sheets that can be copied in 8.5 minutes, multiply 8.5 by the unit rate of 68 to get 578 sheets.

8. D

This question tests the concept of scale. Since the scale is 1 : 16, the actual airplane is 16 times larger than the model airplane. To find the actual length, multiply the length of the model by 16: $38 \times 16 = 608$ inches. Because each of the answer choices is given in feet, convert inches to feet by dividing 608 by 12: $608 \div 12 = 50$ with a remainder of 8, which is equal to 50 feet 8 inches.

9. B

This question tests the concept of similar triangles. When one triangle is similar to another, corresponding sides are in proportion with one another and corresponding angles are equal. Set up a proportion using the known sides of the triangles and the perimeters, lining up corresponding parts. Use the proportion

$$\frac{\text{Side of small triangle}}{\text{side of large triangle}} = \frac{\text{Perimeter of small triangle}}{\text{Perimeter of large triangle}},$$

$$\frac{10\,m}{15\,m} = \frac{x\,m}{34.5\,m}.$$

Cross-multiply the proportion to get $345 = 15x$. Divide each side of the equation by 15 to get $x = 23\,m$.

10. A

This question tests the concept of solving ratio and proportion word problems using variables. Because bagels cost 56 cents each, m bagels cost $56m$ cents. Since n dinner rolls cost as much as m bagels, then divide your known money amount ($56m$) by n (the total amount of rolls) to find out how much one roll costs. Using numerical values for the variables makes problem solving a little easier. Pick any numbers for m and n. Suppose that $n = 4$, and $m = 2$; in other words, 4 dinner rolls cost the same as 2 bagels. Then 2 bagels cost $56 \times 2 = 112$ cents, and one dinner roll would cost $112 \div 4 = 28$ cents. The only answer choice that will result in 28 cents when you plug your chosen values for m and n into the answer choices is choice A: $\frac{(56(2))}{4} = 28$.

RATIOS

> **MATH SPEAK**
>
> A **ratio** is a comparison of two or more different quantities.

Ratios usually appear in fraction form, but there are actually three different ways to write a ratio. A comparison of two different values represented by a and b can be written as $\frac{a}{b}$, $a : b$, or a to b. A ratio can be written using any values, but is considered to be in its most simplified form if there are no common factors between the values and the values in the ratio are integers. Take the ratio 25 to 75. Since each number has a common factor of 25, the ratio can be reduced to 1 to 3 by dividing each number by 25. Therefore, this ratio can be written as 1 to 3, 1 : 3, or $\frac{1}{3}$.

In a more complex comparison such as $2\frac{1}{2}$ to $4\frac{1}{4}$, you can treat the ratio like a division problem in order to simplify. By writing the ratio as a fraction and changing each fraction to improper form, the ratio becomes $\dfrac{2\frac{1}{2}}{4\frac{1}{4}} = \dfrac{\frac{5}{2}}{\frac{17}{4}}$. Since the fraction bar means division, divide $\frac{5}{2}$ by $\frac{17}{4}$. Recall from chapter 3 that dividing by a number is the same as multiplying by its reciprocal: $\frac{5}{2} \times \frac{4}{17} = \frac{20}{34}$. This ratio reduces to $\frac{10}{17}$, 10 : 17, or 10 to 17, when the common factor of 2 is divided out of each number.

Try the following practice question to test your knowledge of equivalent forms of ratios.

Which of the following ratio(s) are equivalent to the ratio $3\frac{1}{4}$ to 13?

    1 to 4   4 to 13   2 to 8   13 to 4   4 to 1

The correct answer to this question is both 1 to 4 and 2 to 8. When the ratio is written as a fraction it becomes $\frac{3\frac{1}{4}}{13}$ which is equal to $\frac{\frac{13}{4}}{13}$ when $3\frac{1}{4}$ is written as an improper fraction. Dividing $\frac{13}{4}$ by 13 is equal to $\frac{13}{4} \times \frac{1}{13}$, which is equal to $\frac{1}{4}$. The correct ratio is 1 to 4, which is also equivalent to 2 to 8 when each number is multiplied by 2.

When using ratios to compare values it is important to realize what type of ratio you are working with. In other words, find out if the ratio compares numbers that represent parts of a larger set, or if the ratio includes a value that represents the entire set. Here's an example:

At a high school, 7 out of every 9 students are involved in extracurricular activities. What is the ratio of students that do not take part in extracurricular activities to the students that do take part in extracurricular activities?

The correct answer to this question is 2 to 7, 2 : 7, or $\frac{2}{7}$. Since 7 out of every 9 students take part in extracurricular activities, the 9 in the ratio represents the whole, or the total number of students at the school. Keep in mind that there are probably more than 9 students enrolled in this school, and that you would have to multiply by a factor to find the actual number, which we are not asked to do for this question. Since $9 - 7 = 2$, then 2 represents the part of the school that does not take part in these activities. The ratio then becomes 2 to 7.

The following problem is an example of a ratio comparing more than two quantities

An artist ordered cans of red paint, white paint, and green paint in the ratio 1 : 2 : 3, respectively. If she ordered a total of 9 cans of green paint, how many cans of white paint did she order?

Even though the ratio involves comparing three different things, the same principles apply that were used when only two quantities were compared. Each number corresponds with the color of paint mentioned in the same order the numbers are listed. That means that the 1 corresponds with the cans of red paint, 2 corresponds with the cans of white paint, and 3 corresponds with the cans of green paint. Since the number in the original ratio that corresponds with green paint is 3 and the total number of cans of green paint is 9, the actual number of cans bought is 3 × the number in the ratio. Therefore, each of the other numbers in the ratio can be multiplied by 3 to find the actual number of cans purchased. Since we are looking for the number of cans of white paint, multiply 2 by 3 to get 6 cans of white paint. You can also use the proportion $\dfrac{\text{ratio of green cans}}{\text{number of green cans}} =$

$\dfrac{\text{ratio of white cans}}{\text{number of white cans.}}$ This proportion would be equal to $\dfrac{3}{9} = \dfrac{2}{x}$.

Since $\dfrac{3}{9} = \dfrac{1}{3}$ and $\dfrac{2}{6} = \dfrac{1}{3}$, then $x = 6$ because each fraction is equal to $\dfrac{1}{3}$.

Rates

A rate is a comparison of two types of units. Some common examples of rates are *miles per hour, feet per second,* and *miles per gallon.* Notice that the key word "per" is often used with rates.

> ### REMEMBER THIS!
> A rate with a denominator of 1 is called a **unit rate.**

For example, if you traveled 100 miles in 2 hours, this would reduce to a unit rate of 50 miles in 1 hour, or 50 miles per hour.

When dealing with rates of speed, it is often useful to use the equation *distance = rate × time* to help you. For instance, if you wanted to find the amount of time it would take to travel 660 miles while going at a rate of 60 miles per hour, you would use the formula and fill in 660 for your distance and 60 for your rate. $660 = 60t$. Divide both sides by 60 to get a time of 11. In other words, it would take 11 hours to travel 660 miles going 60 miles per hour.

Practice 1

Fill in the blank with the correct word, phrase, or symbols. Answers and explanations are located at the end of the chapter.

1. A ratio is a _____ of two or more quantities.

2. The ratio 4 to 10 can also be written as _____, or _____ when expressed in lowest terms.

3. When recognizing problems dealing with rate, very often the key word _____ will be used.

4. If a person can type at the rate of 72 words in 3 minutes, this is equal to the unit rate of _____ words in one minute.

Proportion

> **MATH SPEAK**
> A **proportion** is a comparison of two ratios.

In other words, a proportion is two ratios set equal to each other. In the proportion $\frac{a}{b} = \frac{c}{d}$, *a* and *d* are identified as the *extremes*, and *c* and *b* are identified as the *means*. To solve a proportion, multiply the numerator from the first ratio by the denominator from the second ratio, and then the denominator from the first ratio by the numerator of the second ratio, and set the values equal to each other. In other words, multiply the *means* and set it equal to the product of the *extremes*. After this step, divide each side of this new equation by the number with the variable to get your solution. Note the following example.

Solve for x: $\dfrac{6}{11} = \dfrac{x}{33}$

Using cross-multiplication, the equation becomes $6 \times 33 = x \times 11$. This simplifies to $198 = 11x$. Then divide each side of this new equation by 11. Thus, $x = 18$.

In word problems, proportions are set up so that corresponding units are either one above the other, or directly across from one another. Take the following example.

The ratio of t-shirts to sweaters in a closet is 4 : 5. If there are 12 t-shirts in the closet, how many sweaters are there?

Set up a proportion to solve this question, being careful to line up the units. There are different ways you can approach this problem to get the correct answer. One possible proportion would be $\dfrac{4 \text{ t-shirts}}{5 \text{ sweaters}} = \dfrac{12 \text{ t-shirts}}{x \text{ sweaters}}$. Note that the label of t-shirts appears in the numerators and the label of sweaters appears in the denominators.

Another possible proportion would be $\dfrac{4 \text{ t-shirts}}{12 \text{ t-shirts}} = \dfrac{5 \text{ sweaters}}{x \text{ sweaters}}$, where the numbers corresponding with *t-shirts* appear on the left side and the values corresponding with *sweaters* appear on the right side. Either way, cross-multiplying the means and the extremes results in the equation $4x = 60$. Therefore, $x = 15$. There are 15 sweaters in the closet.

An important concept to note is that many problems that involve ratios and proportions can be categorized as either *part to part* or *part to whole* comparisons. These problems involve different parts of a whole set, and the correct way to set up the proportion to solve the question depends on the question asked. Take, for example, the following problem.

Two numbers are in the ratio 4 : 1. If the sum of the numbers is 30, what is the value of the smaller number?

This is an example of a *part to whole* comparison. since the smaller number (the part) is the intended result and the total (or whole) of the two numbers is given. You want to find the smaller number and are given the sum of the two numbers. Let's look at the given ratio. In the given ratio, the smaller number $= 1$ and the total, or whole, is 5 (since $4 + 1 = 5$). In other words, the *whole* in the ratio is represented by 5 and the *part* that represents the smaller number is represented by 1. Now, set up your 2 ratios so you can cross-multiply to get your solution:

$\dfrac{\text{(smaller number)}}{\text{(whole)}} : \dfrac{part}{whole} = \dfrac{1}{5} = \dfrac{x}{30}$. Cross-multiply to get

$5x = 30$. Divide each side by 5 to get $x = 6$. The smaller number is 6.

In a problem such as the one above, the ratios can also be expressed as the unknown number parts multiplied by x. Then you can write an equation to solve for the values. For instance, the above example can also be solved by writing the equation $4x + 1x = 30$, where $4x$ and $1x$ represent the two unknown number parts and 30 is the known sum. Since the equation can be simplified to $5x = 30$, dividing each side of the equation gives a result of $x = 6$. Therefore, the larger number is $4 \times 6 = 24$ and the smaller number is $1 \times 6 = 6$ ($6 + 24 = 30$).

The following is an example of a *part to part* proportion problem.

In a class, the ratio of students with red hair to black hair is 2 : 3. If there are 6 students with red hair, how many students have black hair in the class?

In this question, parts of the class are being compared (red hair to black hair), so set up a *part to part* proportion $\dfrac{\text{red hair}}{\text{black hair}} = \dfrac{2}{3} = \dfrac{6}{x}$.

Note that both numerators correspond with the number of students with red hair, and the denominators correspond with the number of students with black hair. It is important to use the labels associated with the values to help you set up the proportion correctly. Cross-multiply to get $2x = 18$, and divide each side by 2 to get $x = 9$. There are 9 students with black hair.

Practice 2

Try your hand at the following practice set of proportion questions. For each problem in the left-hand column, match the letter from the right-hand column that corresponds with the correct answer. Answers and explanations are located at the end of the chapter.

5. The ratio of voters who voted for the incumbent candidate versus the challenger was 5 : 2. If there were a total of 35 voters, how many votes did the incumbent get?

A. 20

6. The sum of two numbers is 48. If the numbers are in the ratio 3 : 5, what is the smaller number?

B. 25

7. A committee has 24 women. If the ratio of men to women on the committee is 5 : 6, how many men serve on the committee?

C. 18

Scale

How many times have you opened up a map to try to calculate the distance between two places and used a key such as 1 inch = 1 mile? This is just one example of a **scale**, where a smaller unit is often used to represent a much larger unit. The blueprints of a building or home, model cars, and model airplanes are just a few other examples of the use of scale.

> ### REMEMBER THIS!
> When working with scale models, the **scale** is often given as the ratio *model measurement: actual measurement*.

For example, if the scale of a model airplane is 1 : 25, then the actual airplane would be 25 times as large as the scale model. Therefore, if the length of the model airplane was 2 feet the length of the actual airplane would be $2 \times 25 = 50$ feet in length. Another way to solve this problem is to set up the proportion $\frac{\text{model length}}{\text{actual length}} = \frac{1}{25} = \frac{2}{x}$. Then cross-multiply to get $x = 50$ feet.

As mentioned above, proportions are often helpful when working with scale. Try the following example that deals with a scale on a map:

On a certain real estate map, 1 cm represents $\frac{1}{2}$ km. What distance, in kilometers is represented by $2\frac{1}{2}$ cm?

In order to solve this problem, set up a proportion with the known values in the problem, being careful to line up the units either across from one another or directly above one another. A possible proportion could be $\frac{\text{cm}}{\text{km}} = \frac{1 \text{ cm}}{\frac{1}{2} \text{ km}} = \frac{2\frac{1}{2} \text{ cm}}{x \text{ km}}$

Cross-multiply the values in the proportion to get $1x = \frac{1}{2} \times 2\frac{1}{2}$.

Change to improper fractions to get $x = \frac{1}{2} \times \frac{5}{2}$ which is equal to $\frac{5}{4}$, or $1\frac{1}{4}$ km.

Practice 3

The following question set provides practice in the use of scale. Fill in each answer blank with the expression that best answers the question. Answers and explanations are located at the end of the chapter.

8. If the scale model of a boat measures 6 inches and the model has a scale of 1 : 20, then the actual boat measures _____ feet.

9. The actual height of a train engine is 12 feet. If a scale model of the train is built and is 12 inches long, then the scale used to build the model is _____.

10. The scale on a map is 1 cm = 5 km. If the actual distance between two cites is 37.5 km, then the cities are _____ cm away from each other on the map.

Similarity

> ### MATH SPEAK
>
> When figures have corresponding sides that are in proportion with one another and corresponding angles with the same measure, the figures are **similar**.

Proportions can be used to determine that figures are similar, and also can be used to calculate the missing part or parts of known similar figures. Take, for example, the following diagram of similar triangles. Each set of corresponding angles is congruent, or has the same measure, and each of the known corresponding sides is in proportion to one another.

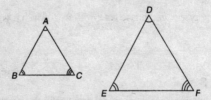

In the above diagram, angle A corresponds with angle D, angle B corresponds with angle E, and angle C corresponds with angle F. In the same manner, side $\overline{AB}$ corresponds with side $\overline{DE}$, side $\overline{BC}$ corresponds with side $\overline{EF}$, and side $\overline{AC}$ corresponds with side $\overline{DF}$.

Now, use a proportion to find the missing side of the larger triangle.

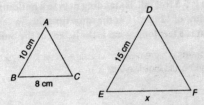

Set up the proportion $\frac{\text{side of smaller}}{\text{side of larger}} = \frac{\text{side of smaller}}{\text{side of larger}}$. Since the side of the smaller triangle that measures 10 cm corresponds with the side of the larger triangle that measures 15 cm, and the side of the smaller triangle that measures 8 cm corresponds with the side of the larger triangle labeled x cm, the proportion becomes $\frac{10}{15} = \frac{8}{x}$. Cross-multiply to get $10x = 120$. Divide each side of the equation by 10 to get $x = 12$ cm. There is another way to look at this problem. Since two corresponding sides were 10 cm and 15 cm, and $\frac{15}{10} = 1.5$, then all sides of the larger triangle are 1.5 times larger than all sides of the smaller triangle. Multiply 8 cm by 1.5 to get the corresponding measurement of 12 cm. These 2 triangles are in the scale ratio 1 : 1.5.

> ### REMEMBER THIS!
> The ratio of the sides of two similar figures will be the same as the ratio of the perimeters of the same two similar figures.

For example, if the sides of two similar triangles are in the ratio 1 : 3, then the perimeters of the triangles will also be in the ratio of 1 : 3.

An additional place where similarity and scale are often used is in shadow problems. In this type of problem, you can use indirect measurement to find the height of something that may be too tall or too large to measure yourself. Take, for example, the following scenario:

Justin, who is 5.5 feet tall, is standing next to a building that casts a shadow of 12 feet. If, at the same time, Justin casts a shadow that is 3 feet long, what is the height of the building?

Drawing diagrams is a great way to visualize scenarios. A picture of the above scenario could look like this:

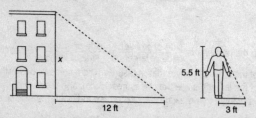

Notice the similar triangles that appear once the diagram is drawn; the corresponding angles are congruent and the corresponding sides are in proportion. To solve this problem, set up a proportion like the ones used for similar triangles. Line up the corresponding parts/labels as in the proportion $\dfrac{\text{height of person}}{\text{shadow of person}} = \dfrac{\text{height of building}}{\text{shadow of building}}$.

The proportion then becomes $\dfrac{5.5}{3} = \dfrac{x}{12}$. Cross-multiply to get the equation $66 = 3x$; divide each side of the equation by 3 to get $x = 22$. The building is 22 feet tall.

RATIO AND PROPORTION WORD PROBLEMS USING VARIABLES

Many times a question or answer choices on a test may contain a series of variables that make the question seem more difficult than it really is. To solve this type of question, simply use labels just as you would in a question that contained numbers. The following is an example of this strategy.

Celine can paint n tiles in m minutes. Working at this rate, how many tiles can she paint in one hour?

Since you are not given any numerical values, your answer will have to include the variables m and n. Set up a proportion as if you had numerical values. Rate is found by dividing the total number of tiles by the number of minutes, or $\dfrac{n \text{ tiles}}{m \text{ minutes}}$. Since there are 60 minutes in 1 hour (your given value), set up the proportion $\dfrac{n \text{ tiles}}{m \text{ minutes}} = \dfrac{x \text{ tiles}}{60 \text{ minutes}}$. Cross-multiply to get the equation $60n = mx$. Since you are looking for x (the question asks for number of tiles), divide each side of the equation by m to get $x = \dfrac{60\,n}{m}$. This equation is your correct answer.

Another strategy to use in this type of question is to replace the letters with numbers of your choice and try evaluating to make sure you have the correct variable expression. Suppose Celine can paint 10 tiles in 2 minutes, which is a rate of 5 tiles in 1 minute. If she can paint 5 tiles in 1 minute, then she can paint $5 \times 60 = 300$ tiles in 1 hour. If you are checking to see if your variable expression is correct, then plug the numerical values into the expression: $\dfrac{60\,n}{m}$ would simplify to

$$\dfrac{60 \times 10}{2} = \dfrac{600}{2} = 300 \text{ tiles in 1 hour, which is the correct solution.}$$

Practice 4

The following true/false questions ask you to use the skills just presented on similar figures and word problems using variables. Remember, repetition is the key to mastery. Answers and explanations are located at the end of the chapter.

11. **T   F**   In the figure below, the length of side $\overline{AB}$ would be equal to 11 cm.

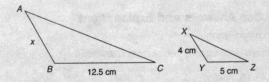

12. **T   F**   If the perimeters of two similar triangles are 33 inches and 42 inches, respectively, and the measure of the shortest side of the smaller triangle was 11 inches, then the shortest side of the larger triangle would measure 14 inches.

13. **T   F**   A 24-foot flagpole casts a shadow that is 16 feet long. At the same time, a person 6 feet tall would cast a shadow that is 4 feet long.

14. **T   F**   If the cost of 3 notebooks is equal to d dollars, then the cost of f notebooks can be represented by the expression $\frac{3f}{d}$.

SUMMARY

During this chapter, you reviewed and practiced many skills and concepts pertaining to the important principles of ratio, rates, and proportions. Here are five important things about ratio and proportion that you should take away from this lesson:

- A ratio is a comparison of different quantities.
- A rate is a comparison of different units and uses the key word "per."
- A unit rate is a rate with a denominator of 1.
- A proportion is a comparison of two ratios; some are part to part, some are part to whole. Use cross-multiplication to solve proportions.
- You can use proportions to help you solve real-world applications of ratios, such as scale, similarity, and various word problem situations.

Practice Answers and Explanations

1. **comparison**

A ratio is a comparison of two or more quantities.

2. $4:10, \frac{2}{5}$

3. per

The key word often used with rates is the word "per," as in miles

per hour, price per gallon, etc.

4. 24 words in 1 minute

A rate of 72 words in 3 minutes can be simplified to 24 words in
1 minute by dividing 72 by 3.

5. B

Set up the *part to whole* proportion $\frac{5}{7} = \frac{x}{35}$, and then cross-multiply

and divide to get $x = 25$. Or, set up the equation $5x + 2x = 35$, which

turns into $7x = 35$. Divide both sides by 7 to solve find that $x = 5$. Then
multiply 5×5 (the larger number) to get the same answer of 25.

6. C

Set up the part to whole proportion using the given smaller number:

$\frac{3}{8} = \frac{x}{48}$. Then, cross-multiply and divide to get $x = 18$. Or, set up the

equation $5x + 3x = 48$. $8x = 48$. $x = 6$. Multiply 6×3 (the smaller
number) to get the same answer of 18.

7. A

Set up the part to part proportion and cross-multiply and divide to get
$x = 20$.

8. 10 feet

First, set up a proportion using the scale and the length of the boat:

$\frac{1}{20} = \frac{6}{x}$. Cross-multiply to get $x = 120$ in. Be careful to note what

units the question is asking for. Divide by 12 to find the length in feet.
120 divided by 12 equals 10 feet.

9. 1 : 12

Be careful to pay attention to the units being used. Since 12 inches is
equal to 1 foot, use feet when comparing. The scale can be expressed
as the model measurement : actual measurement. Since the model is
1 foot long and the actual is 12 feet long, the scale becomes 1 : 12.

10. 7.5 cm

Use a proportion and line up the corresponding units.

$\frac{1 \text{ cm}}{5 \text{ km}} = \frac{x \text{ cm}}{37.5 \text{ km}}$. Cross-multiply to get $5x = 37.5$. Divide each side by 5 to get $x = 7.5$. They are about 7.5 cm apart on the map.

11. False

By setting up the corresponding sides in a proportion you get $\frac{\text{side of large}}{\text{side of small}}$ which is equal to $\frac{x}{y} = \frac{12.5}{5}$. So $\frac{12.5}{5} = \frac{x}{4}$.
Cross-multiply to get $5x = 50$, and divide each side of the equation by 5 to get $x = 10$. Side $\overline{AB}$ is 10 m, not 11 m as stated.

12. True

Set up the proportion side of $\frac{\text{side of small}}{\text{perimeter of small}} = \frac{\text{side of large}}{\text{perimeter of large}}$, which is equal to $\frac{11}{33} = \frac{x}{42}$. Cross-multiply and then divide each side of the equation by 33 to get $x = 14$ inches.

13. True

Set up the proportion $\frac{\text{height of person}}{\text{shadow of person}} = \frac{\text{height of flagpole}}{\text{shadow of flagpole}}$. The proportion would be $\frac{6}{x} = \frac{24}{16}$. Cross-multiply to get $96 = 24x$. Divide each side by 24 to get $x = 4$. The shadow of the person would be 4 feet long.

14. False

The expression should be $\frac{df}{3}$. Since the cost of 3 notebooks is d dollars, then $3 \times n$ notebooks $= d$ dollars. Divide each side of this equation by 3 to get $n = \frac{d}{3}$. This represents the cost of 1 notebook. Multiply this expression by f to get the cost of f notebooks: $f \times \frac{d}{3} = \frac{df}{3}$.

CHAPTER 4 TEST

Try the following questions to see what you have learned about the concepts of ratio and proportion. Following the questions are complete answer explanations to help you assess your understanding.

1. A ratio is in reduced form if there are no common factors between the numbers in the ratio, and the numbers are
   (A) irrational   (B) operations   (C) integers
   (D) decimals   (E) fractions

2. Which of the following is NOT equivalent to the ratio 8 to 6?
   (A) $\frac{4}{3}$ (B) $\frac{6}{8}$ (C) $4:3$ (D) $\frac{16}{12}$ (E) $32:24$

3. The ratio of dogs to cats in a show is 9 to 11. If the show only allows dogs and cats, what is the ratio of cats to the total animals at the show?
   (A) 9 to 11   (B) 9 to 20   (C) 11 to 9
   (D) 11 to 20   (E) 20 to 9

4. Solve for x: $\dfrac{17}{x} = \dfrac{51}{6}$
   (A) 2.8   (B) 2   (C) 3   (D) 5.6   (E) 51

5. After completing $\dfrac{7}{10}$ of his math homework assignment, Josh has 15 more questions to complete. What is the total number of questions in this assignment?
   (A) 17   (B) 21   (C) 25   (D) 32   (E) 50

6. In a certain county, 2 out of 3 households have 2 or more telephones. If there are approximately 210,000 households in the county, how many have less than 2 telephones?
   (A) 7,000   (B) 14,000   (C) 21,000
   (D) 70,000   (E) 140,000

7. Steven types at a rate of 44 words per minute. At this rate, how many words can he type in 8 minutes?

   (A) 352   (B) 5.5   (C) 6   (D) 176   (E) 358

8. Chelsea feeds her dog 32 ounces of dog food once a day. At this rate, how many days will a 40-pound bag of dog food last?

   (A) 20   (B) 24   (C) 32   (D) 40   (E) 48

9. The model of a car is built in the ratio 1 : 40. If the actual length of the car is 10 feet, what is the length of the model, in inches?

   (A) 3 in   (B) 4 in.   (C) 10 in.   (D) 40 in.   (E) 400 in.

10. On a road map, the distance between two cities is approximately 14 inches. If the actual mileage between the two cites is 70 miles, then what is the scale used on the map?

    (A) 1 mile = 5 inches   (B) 1 inch = 14 miles

    (C) 1 inch = 5 miles   (D) 1 mile = 70 inches

    (E) 1 inch = 70 miles

11. A scale replica of a building is 18 inches tall. If the actual height of the building is 36 feet, then the scale used between the model and the actual building is 1 inch = _____.

    (A) 18 inches   (B) 1.5 feet   (C) 2 feet

    (D) 18 feet   (E) 24 feet

12. In the diagram below, triangle JKL is similar to triangle MNO. What is the measure of side $\overline{MO}$?

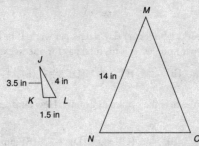

(A) 7 inches   (B) 10.5 inches   (C) 14.5 inches

(D) 16 inches   (E) 9 inches

13. A photographer wishes to enlarge a picture that is $2\frac{1}{2}$ inches wide by 3 inches long to have a new width of 10 inches. What will be the length of the enlargement?

(A) 3 inches   (B) 6 inches   (C) 8 inches

(D) 10 inches   (E) 12 inches

14. At the same time, a tree casts a shadow that is 5 feet long and a person casts a shadow that is 2.5 feet long. If the tree is 12 feet tall, how tall is the person?

(A) 4.5 feet   (B) 6 feet   (C) 6.5 feet

(D) 7 feet   (E) 12.5 feet

15. The length, in meters, of the sides of a quadrilateral are 2, 3, 4, and 4.5. What is the length of the shortest side of a similar quadrilateral if its perimeter is 40.5 meters?

(A) 5 meters   (B) 6 meters   (C) 10 meters

(D) 12 meters   (E) 20 meters

16. If the cost of r pencils is t cents, what is the cost of x pencils?

   (A) $\frac{tx}{r}$ (B) $\frac{r}{tx}$ (C) $\frac{t}{rx}$ (D) $\frac{rt}{x}$ (E) $\frac{xr}{t}$

17. Peter works h hours per day, d days per week. If he earns c dollars per hour, what is his total pay for one week?

   (A) cdh (B) $\frac{cd}{h}$ (C) $\frac{cd}{d}$ (D) $\frac{d}{ch}$ (E) $\frac{c}{dh}$

18. Todd rides his bike x miles in y hours. How far does he ride in z hours?

   (A) $\frac{xz}{y}$ (B) $\frac{x}{yz}$ (C) xyz (D) $\frac{2}{xy}$ (E) none of these

Answers and Explanations

1. C

By definition, a ratio is in reduced, or simplest form, if there are no common factors between the numbers in the ratio and the numbers in the ratio are integers.

2. B

Be careful about the order of the numbers used in a ratio. The ratio 8 to 6 can be written in a variety of equivalent forms, including answer choice (A), choice (C), choice (D), and choice (E). The only answer choice that is not equivalent is $\frac{6}{8}$.

3. D

Because the label *dogs* is mentioned first in the sentence, the first number listed corresponds with the number of dogs in the show. Thus, 9 corresponds with the number of dogs and 11 corresponds with the number of cats. The total number of animals would then be represented by the value $9 + 11 = 20$. Since the question asks for the ratio of cats to the total number of animals in the show, the final ratio would then be 11 to 20.

4. B

To solve this proportion, cross-multiply the *means* and the *extremes* and set them equal to each other. Then divide to solve the equation.

Cross-multiplying gives the equation $102 = 51x$. Divide each side by 51 to get $x = 2$.

5. E

Set up a proportion to solve this question. Since Josh has completed $\frac{7}{10}$ of the assignment, 7 represents the part he has finished and 10 represents the whole assignment. This makes $10 - 7 = 3$ represents the part he still needs to complete. Since he still has 15 questions to complete, then set up the proportion $\frac{\text{part}}{\text{whole}} = \frac{3}{10} = \frac{15}{x}$. Cross-multiply to get $3x = 150$. Divide each side by 3 to get $x = 50$. There were a total of 50 questions on his assignment.

6. D

Because 2 out of 3 households have 2 or more telephones, then 1 out of every 3 households will have less than 2 telephones. Set up the proportion $\frac{\text{part}}{\text{whole}} = \frac{1}{3} = \frac{x}{210,000}$, and cross-multiply to get $210,000 = 3x$. Divide each side of the equation by 3 to get $x = 70,000$. 70,000 households in this county have less than 2 telephones.

7. A

Since Steven types at a rate of 44 words per 1 minute, multiply 44 times 8 to get 352 words in 8 minutes. Another way to solve this problem is to set up the proportion $\frac{\text{words}}{\text{minute}} = \frac{44 \text{ words}}{1 \text{ minute}} = \frac{x \text{ words}}{8 \text{ minutes}}$. Cross-multiply to get $352 = x$.

8. A

Look closely at the units your problem asks for. First, convert 40 pounds to ounces. Since there are 16 ounces in 1 pound, multiply 40 by 16 to get 640 ounces. She uses 32 ounces per day, so divide the total number of ounces by the ounces per day to figure out how many days the bag will last. $\frac{640 \text{ ounces}}{32 \text{ ounces per day}} = 20$ days.

9. A

Since the ratio of the model to the actual car is 1 : 40, set up the

proportion $\frac{\text{model}}{\text{actual}} = \frac{1}{40} = \frac{x}{10}$. Cross-multiply to get $10 = 40x$.

Divide each side of the equation by 40 to get 0.25 feet. To convert to inches, multiply 0.25 by 12 inches to get 3 inches.

10. C

Set up the proportion $\frac{1 \text{ inch}}{14 \text{ inch}} = \frac{x \text{ miles}}{70 \text{ miles}}$. Cross-multiply to get the

equation $70 = 14x$. Divide each side of the equation by 14 to get $x = 5$. Therefore, 1 inch is equal to 5 miles.

11. C

Set up the proportion $\frac{\text{model}}{\text{actual}} = \frac{1}{x} = \frac{18 \text{ inches}}{36 \text{ feet}}$. Cross-multiply to get

$36 = 18x$. Divide each side of the equation by 18 to get $x = 2$ feet.

12. D

Set up the proportion $\frac{\text{side of small}}{\text{side of large}} = \frac{\text{side of small}}{\text{side of large}}$. The proportion

becomes $\frac{3.5}{14} = \frac{4}{x}$, where x represents the length of side . Cross-

multiply to get $3.5x = 56$. Divide each side of the equal sign by 3.5 to get $x = 16$.

13. E

Set up the proportion $\frac{\text{width}}{\text{length}} = \frac{2\frac{1}{2}}{3} = \frac{10}{x}$. Cross-multiply to get

$2.5x = 30$. Divide each side of the equation by $2\frac{1}{2}$ or 2.5 to get $x = 12$ inches.

14. B

Set up the proportion $\frac{\text{height of object}}{\text{length of shadow}} = \frac{12}{5} = \frac{x}{2.5}$ Cross-multiply

to get $30 = 5x$. Divide each side of the equation by 5 to get $x = 6$. The person is 6 feet tall.

15. B

Note that the perimeter of the first quadrilateral is the sum of the sides: $2 + 3 + 4 + 4.5 = 13.5$ meters. Set up the proportion length of

$$\frac{\text{shortest side}}{\text{perimeter of object}} = \frac{2}{13.5} = \frac{x}{40.5}.$$ Cross-multiply to get

$81 = 13.5x$. Divide each side of the equation by 13.5 to get $x = 6$. The length of the shortest side of the second quadrilateral is 6 meters.

16. A

Since the cost of r pencils is t cents, then the cost of one pencil is

$\frac{t \text{ cents}}{r \text{ pencils}}$ or $\frac{t}{r}$. Therefore, the cost of x pencils would be $\frac{t}{r} \times x$ which

is equal to $\frac{tx}{r}$. Another approach to this question is to substitute values

for the variables and see which answer choice matches the evaluated solution. Suppose that $r = 5$ pencils, $t = 50$ cents, and $x = 6$ pencils. Therefore, 5 pencils cost 50 cents, so each pencil would cost $50 \div 5 = 10$ cents. Then the cost of 6 pencils would be $10 \times 6 = 60$ cents. The answer choice that would also have a value of 60 cents, with these same values plugged into the variables, is answer choice A.

$$\frac{((50)(6))}{(5)} = 60$$

17. A

Since Peter works h hours per day, d days per week, he works $h \times d$, or dh, hours per week. If he makes c dollars per hour, multiply this number times his total number of hours, or $c \times dh$. The expression becomes cdh. Another approach to this problem is to substitute in values for the variables and see which answer choice would also result in the evaluated result. Suppose that $h = 8$ hours, $d = 5$ days, and $c = 10$ dollars per hour. Then Peter works 8 hours per day, 5 days a week, or $8 \times 5 = 40$ hours per week. If he then makes \$10 per hour, his total pay for the week is $40 \times \$10 = \$400$. The answer choice that would also have a value of \$400 with these same values plugged into the variables is answer choice A.

18. A

Solve the problem by using values and labels for each of the variables in order to help find the expression. Use the formula *distance = rate × time*, which can also be *written as rate* $= \frac{distance}{time}$. Assume that Todd rides his bike 25 miles in 5 hours. This would then mean that Todd rides his bike $\frac{25 \text{ miles}}{5 \text{ hours}} = 5$ miles per hour. Since you divided the total miles by the time, the first part of the expression becomes $\frac{x \text{ miles}}{y \text{ hours}}$. Now set up the proportion $\frac{x \text{ miles}}{y \text{ hours}} = \frac{? \text{ miles}}{z \text{ hours}}$. Cross-multiply to get $xz = y \times ?$. Divide each side of the equation by y to get the expression $\frac{xz}{y}$.

Percent

BUILDING BLOCK QUIZ

The previous chapter dealt with the concepts of ratio and proportion. A percent is a special kind of ratio that compares a quantity to the base number of 100. You need a chapter just on this special ratio alone, because you encounter percents in so many aspects of daily life. Start off with this Building Block Quiz as an introduction. Remember, the first three questions will review concepts from last chapter.

1. Solve for x: $\dfrac{7}{50} = \dfrac{x}{75}$

   (A) 10.5 (B) 10 (C) 14 (D) 3.5 (E) 5.75

2. If triangle ABC is similar to triangle DEF, what is the measure of side EF?

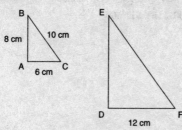

   (A) 16 (B) 20 (C) 4 (D) 5 (E) 14

3. The ratio of boys to girls on the debate team is 2 to 3. If there are 25 members of the debate team, how many members are girls?

    (A) 10   (B) 5   (C) 15   (D) 20   (E) 3

4. Change $\frac{2}{5}$ to a percent.

    (A) 25%   (B) 40%   (C) 80%   (D) 2.5%   (E) 250%

5. What is 35% of 300?

    (A) 35   (B) 300.35   (C) 105   (D) 8.5   (E) 70

6. A washing machine originally priced at $599.00 is on sale for 40% off. What is the total price including sales tax? Assume a sales tax percentage of 7%.

    (A) $512.74   (B) $281.53   (C) $401.33
    (D) $384.56   (E) $359.40

7. Hannah deposits $1,250 in a savings account that pays 5% simple interest. How much interest will be earned after 2 years?

    (A) $62.50   (B) $625.00   (C) $12.50
    (D) $1,250.00   (E) $125.00

8. Given the grades below for a history examination, what percentage of the students had a passing grade of 65 or higher?

Grade Range	Number of Students
90-100	60
80-89	70
70-79	60
65-69	50
60-64	10
below 60	50

(A) 80%   (B) 240%   (C) 16.7%
(D) 50%   (E) 20%

9. The realtor gains a 3% commission on any home sale. What is the commission on a home selling for $89,000?
   (A) $267.00   (B) $2,670.00   (C) $86,330.00
   (D) $300.00   (E) $8,900.00

10. The number of workers using public transportation rose from 5,700 to 7,200 people. What is the percent increase?
    (A) 1,500   (B) 150   (C) 15   (D) 20.8   (E) 26.3

Answers and Explanations

1.   A

This problem is a review from the last chapter on cross-multiplication. Cross multiply to set up the equation $50x = 7 \times 75$, or $50x = 525$. Divide both sides by 50 to get $x = 10.5$.

2.   B

This problem reviews the concept of similarity. For similar triangles, the sides are in proportion. Side AC of length 6 corresponds to side DF of length 12. Side BC corresponds to side EF. Set up a cross product $\frac{AC}{BC} = \frac{DF}{EF}$, or $\frac{6}{10} = \frac{12}{x}$. Cross multiply to get $6x = 120$. Divide both sides by 6 to get $EF = 20$.

3. C

This problem tests your understanding of ratios. The ratio of boys to girls is given as 2 to 3. The ratio of girls to all is 3 to 5, or $\frac{3}{5}$. Set this equal to the comparable part-whole ratio of $\frac{x}{25}$. Cross multiply to get $5x = 75$. $x = 15$ girls.

4. B

This question assesses your knowledge of the conversion of a fraction to a percent. You can change $\frac{2}{5}$ to a percent by setting up the equivalence $\frac{2}{5} = \frac{x}{100}$. Cross multiply to get $5x = 200$. Divide both sides by 5 to get 40, or 40%.

5. C

To solve this percent of a number problem, first convert 35% to a decimal. $35\% = 0.35$. Then multiply this decimal by 300 to find 35% of 300. 0.35 times 300 is 105.

6. D

This is an application of percentages. First, find the sales price and then add on the sales tax. There is a discount of 40 percent off the price, so you will pay 60 percent of the original price. Sixty percent of $599.00 is $0.60 \times 599 = 359.40$. Now, calculate 7 percent of this price to find the sales tax which will be added onto the price to be paid. The sales tax is $0.07 \times 359.40 = 25.16$. The total price is $359.40 + 25.16 = \$384.56$.

7. E

This problem assesses your knowledge of the simple interest. Use the simple interest formula: $I = prt$, where I is the interest earned, p is the principle amount invested, r is the percentage rate, written as a decimal, and t is the time in years. $I = 1,250 \times 0.05 \times 2 = \$125.00$.

8. A

This question deals with the percentage relationship of part-whole. The percent that passed is the amount of students with passing grade divided by the total number of students who took the test. The number that passed is $60+70+60+50=240$ students. The total is 240 plus the number that failed, or 240+10+50=300 students. The percentage is 240 divided by 300, or 0.8 which is 80 percent. If you chose (B), you found the number of students who had a passing grade, not the percentage. Choice (C) is the percentage of students who received a grade between 65 and 69. Choice (D) is just the number of students who received a grade between 65 and 69. Choice (E) is the percentage of students who had a failing grade.

9. B

This problem tests your knowledge of sales commissions. The commission is 3 percent, or 0.03 of the selling price of $89,000. Multiply 0.03 times 89,000 to get $2,670.00.

10. E

This problem tests your knowledge of percent increase. To calculate this percent, first find the change in number of workers. $7,200 - 5,700$ is 1,500. Set up the cross product $\frac{percent}{100} = \frac{change}{original\ amount}$, or $\frac{x}{100} = \frac{1,500}{5,700}$. Cross multiply to get $5,700x = 150,000$. Divide both sides by 5,700 to get $x = 26.3$, that is 26.3 percent.

WORKING WITH PERCENTS

Percents are used often in everyday life. Any time you shop for a sale, you work with a percentage off of the original price of merchandise. Many institutions give test results as percentages. Budget reports often break down expenses or revenues based on percentages. Health conscious consumers are concerned with the percentage of fats or carbohydrates in various foods. You can solve percent problems with methods similar to those described in chapter 4. This chaper will also introduce you to additional methods for solving.

Converting between Decimals and Percents

> ### MATH SPEAK
>
> A **percent** is a special ratio that compares a numerical quantity to 100.

Forty-five percent (45%) is the same as $\frac{45}{100}$, and also the same as 0.45. Because of this fact, it is easy to convert a percent number to a decimal.

> ### REMEMBER THIS!
>
> To change a percent to a decimal, remove the percent symbol (%) and divide by 100.
>
> To change a decimal to a percent, just multiply by 100.

Because of our place value number system, to divide by 100 you just have to move the decimal point two places to the left and remove the percent symbol (%). For example: 56% is 0.56, 230% is 2.30, and 4% is 0.04.

To multiply any number by 100, simply move the decimal point two places to the right and add the percent symbol %. For example: 0.76 is 76%, 1.34 is 134%, and 0.06 is 6%.

Converting between Fractions and Percents

Remember that a percent is a ratio, or fraction, that compares a numerical quantity to 100. To convert from a fraction $\frac{a}{b}$ to a percent you can use a variable to represent the unknown percent and set up a proportion.

REMEMBER THIS!

To change a fraction to a percent use the proportion $\dfrac{x}{100} = \dfrac{a}{b}$ and cross multiply to solve for the variable x.

For example, to change $\dfrac{4}{5}$ to a percent, set up the proportion

$\dfrac{x}{100} = \dfrac{4}{5}$. Cross multiply to get $5x = 4 \times 100$, or $5x = 400$. Divide

both sides by 5 to get $x = 80$, or 80%.

REMEMBER THIS!

To convert from a percent into a fraction, just put the given percent over 100, remove the percent symbol (%), and simplify if needed.

For example, to convert 68% to a fraction, simply place the given

percent over 100: $\dfrac{68}{100}$. Divide the numerator and denominator by 4

to simplify the fraction: $\dfrac{68}{100} \div \dfrac{4}{4} = \dfrac{17}{25}$. If the percent is a decimal,

such as 43.2%, the fraction will be $\dfrac{43.2}{100}$. First, you must multiply the

fraction by $\dfrac{10}{10}$ to clear the decimal point from the numerator:

$\dfrac{43.2}{100} \times \dfrac{10}{10} = \dfrac{432}{1,000}$. Now, simplify the fraction by dividing the

numerator and the denominator by 8, the greatest common factor.

Therefore, $43.2\% = \dfrac{432}{1,000} \div \dfrac{8}{8} = \dfrac{54}{125}$.

Practice 1

In the following exercise, match each item in the left column with its equivalent value in the right column. Answers are located at the end of the chapter:

1. 13%   A. $\dfrac{3}{10}$

2. 30%   B. 0.13

3. 1.3%   C. $\dfrac{8}{10}$

4. .8%   D. $\dfrac{8}{1,000}$

5. .08%   E. 0.013

6. 80%   F. 0.0008

The Percent of a Number

You can think of a percent as a part-whole relationship. If you are asked to find the percent of a number, you are trying to find the part of the whole number that is represented by the given percent.

> ### REMEMBER THIS!
> The key word "of" in mathematics usually means multiply.
> The key word "is" in mathematics means equals.

The equation "part is percent times number" can be used to solve problems of this type. For example, to find 18% of 250, change 18% to a decimal and then multiply by 250: $0.18 \times 250 = 45$.

You may be asked to find a missing percent, such as *252 is what percent of 600?* Set up a simple equation, substituting an equal sign for "is" and a multiplication sign for "of": $252 = 600x$. Divide both sides by 600 to get 0.42, or 42%. Note that when using the equation method the percent is expressed as a decimal number.

Another type of problem may give you the part and the percent, and ask you to find the whole number. For example: *26% of some number is 105.3. What is the number?* Again, set up a simple equation: $0.26x = 105.3$. Divide both sides by 0.26 to get $x = 405$.

An alternate method that you may prefer is to set up a proportion $\frac{part}{whole} = \frac{\%}{100}$ and cross multiply to find the missing term. The part is associated with the word "is" and the whole is associated with the word "of". For example, if asked *What percent of 600 is 114?*, set up the proportion $\frac{114}{600} = \frac{x}{100}$. Cross multiply to get $600x = 11,400$. Divide both sides by 600 to get 19, which is 19%. As you can see, when using the proportion method, the percent is not expressed as a decimal.

Practice 2

The key to mastering percents is practice! Use the following bank of answers to solve the problems below. Answers and explanations are located at the end of the chapter.

65%	45	65	30
85%	30%	450	.85%

7. What is 26% of 250?

8. 629 is what percent of 740?

9. 11.25 is 25% of what number?

10. What is 200% of 15?

There are many examples of the percent relationship of part to whole. For example, a company's expense budget may be represented in the following circle graph:

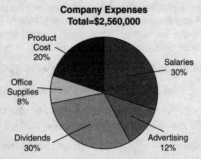

Company Expenses
Total=$2,560,000

Product Cost 20%

Salaries 30%

Office Supplies 8%

Dividends 30%

Advertising 12%

The "whole" is the total expenses. Each slice of the circle graph is a part of the total. To find the amount of money allocated to advertising, use the percent value (12%) and total budget allocation given in the graph (2,560,000), and set up a proportion: $\frac{12}{100} = \frac{x}{2,560,000}$. Cross multiply to get: $100x = 30,720,000$. Divide both sides by 100 to get the monetary expense related to advertising: $307,200.

Another example is the amount of female students at a given college of 3,500 students. If the college brochure states that 36% of the students are female, then the whole is the total number of students (3,500) and the part is the number of female students (36%). Find the amount of female students by multiplying 0.36 by 3,500 to get 1,260 females.

Sometimes a salesman earns a percentage commission off of the sale of items. The commission is a percentage of the selling price — one part of the whole sale price. Find the monetary amount of this commission by multiplying the percent, written as a decimal, by the selling price.

For example, Rita earns a 5% commission on each pair of shoes she sells. If she sells 10 pairs of $15.00 shoes each day, how much commission will she earn after 5 working days?

At this rate, Rita will sell 50 pairs of shoes (10 each day for 5 working days) at $15.00 each, for a total in sales of $(50)(15) = \$750$. Her commission is 5% of this amount, or $(0.05)(750) = \$37.50$.

Try the following percentage word problems to test your understanding. Answers and explanations are located at the end of the chapter.

Practice 3

11. The standard tip percentage for a waitress is 15%. What is the tip amount on a meal total of $32.00?

12. Joe earns a 3% commission for every home he sells. What is his commission on a $132,000.00 home that he sells?

13. Of the 50,000 commuters to the city, 60% of them use public transportation. How many commuters use public transportation?

14. In the following chart, what is the percentage of people who wear a shoe size of 9 or larger?

Shoe Size	Number of People
6	10
7	15
8	25
9	20
10	10

PERCENT APPLICATIONS

Percent Increase or Decrease

There are times that you may be interested in how much an amount has changed, either as an increase or a decrease. Stock price values are examined in this manner. Other applications include a company that tracks a change in sales or fixed costs or a club that tracks a change in attendance at meetings. The government and news organizations keep a close watch on the change in the number of unemployed or uninsured citizens. All of these examples could be solved as percent increase or decrease applications.

REMEMBER THIS!

Percent increase or decrease is calculated as

$$\frac{\text{change}}{\text{original amount}} = \frac{\text{percent}}{100}.$$

It is important to realize that the part is the change that occurs, and the whole is the starting amount. The new amount does not enter into the formula to calculate the percent. The new amount is only used to calculate the change from the original. For example, to find the percent decrease in the town board meeting in which the attendance went from 125 people down to 50 people, first calculate the change in attendance, $125 - 50 = 75$. Set up the proportion $\frac{75}{125} = \frac{x}{100}$. Cross multiply to get $125x = 7,500$. Divide both sides by 125 to get $x = 60$, or 60%. You can also solve percent increase or decrease problems by setting up an equation such as *change = percent × original*. The above example would be set up as $75 = 125x$. For this equation, you divide both sides by 125 to get $x = 0.60$, which is also 60%.

Practice 4

Use the percentage bank of answers to find the missing percents. (Percents are rounded to the nearest tenth). Answers and explanations are located at the end of the chapter.

| 60% | 14% | 56% | 25% | 5.6% | 4% |
| 16.7% | 5% | 20% | 10.3% | 24% | |

15. What is the percent increase from 25 to 39?

16. What is the percent decrease from 90 to 85?

17. What is the percent decrease from 24 to 20?

18. What is the percent increase from 125 to 200?

19. The stock price at Buyer's Choice rose from $12.50 to $15.00 per share. What is the percentage increase?

20. The total number of teams at a weekly trivia competition went from 25 to 19 due to a location change. What is the percentage decrease?

Sales and Sales Tax

The retail world of buying and selling is filled with examples of percentage use. When there is a sale on merchandise, the discount is represented as a percent decrease, where the change in price is the amount of money saved. When merchandise is sold there is usually a sales tax involved that is calculated as a percent —a percent increase to the original price.

For example, if you purchase a computer for 35% off the retail price of $1,000.00. What is the sale price?

The discount, (change in price), equals 0.35 times the original price, or discount = (0.35)(1,000). Multiply to get the discount of $350.00. This is the change in price, so the sale price is 1,000 − 350 = $650.00.

Hockey sticks originally sell for $129.99. They are on sale for 20% off this price. There is also a 6% sales tax. What will be the total price paid?

This is a two-step problem. First, you can calculate the sale price, and then calculate the tax, which will be added to the sale price. The discount is (0.20)(129.99) = $26.00, rounded to the nearest penny. The sale price is thus 129.99 − 26.00 = $103.99. For step two, calculate the tax on the sale price. The tax is (0.06)(103.99) = $6.24. The total price paid is the sale price plus the tax, or 103.99 + 6.24 = $110.23.

There is an alternate, shortcut method to calculating sale prices. If you think of the original price as 100%, then the percentage savings is one part of the original price, and the price paid is the other part of the original price. Therefore, if an item is 35% off of the original price, you will pay 65% of the original price. If an item is 20% off, you will pay 80% of the original price. Likewise, to find the price that you will pay with the sales tax percentage increase, first understand that you will pay the sale price plus the sales tax percentage. If the sales tax is 7%, then you will pay 107% of the sale price to the cashier; if the sales tax is 4%, you will pay 104% of the sale price.

For example, to find the total price to pay for a dress originally priced at $70.00 on sale for 30% off, set up an equation that reflects that the sale price is 70% of $70.00. Change 70% to a decimal and then multiply: $0.70 \times \$70.00 = \$49.00$. If there is then a 6% sales tax, you will pay the cashier 106% of this sale price, or $1.06 \times \$49.00 = \$51.94$.

This method is especially helpful if you are given the sales price and the sale percentage discount, and you want to calculate the original price of an item. For example, if a barbeque grill is sold for $259.35, and this price reflects a 35% discount, what was the original price of the grill? The sale price of the grill is 65% of the original price. Set up an equation: $259.35 = 0.65x$, where x represents the original price. Divide both sides of the equation by 0.65 to get $399.00.

Practice 5

Practice is essential for understanding mathematics. Try the following sales tax exercises. Answers and explanations are located at the end of the chapter.

21. What is the sales tax on a $289.00 portrait purchase, if the sales tax percentage is 7%?

22. Lara purchases a new coat for 30% off of the original price of $160.00. What is the sale price?

23. A camera, originally priced at $350.00, is on sale for 25% off of the price. There is a 4% sales tax. What will be the purchase price, including sales tax?

24. Josh pays $3,450.00 for a piano. This price reflects a sale percentage discount of 20%. What was the original price of the piano?

Simple Interest Rates

Banks use interest rates in the form of percentages to either award interest to savings accounts or to charge interest on loans. Interest rates are used so often that there is a formula used to calculate their values.

REMEMBER THIS!

Simple interest is calculated with the formula $I = prt$, where I is the interest charged or paid out, p is the principle amount that is saved or borrowed, r is the percentage rate written as a decimal, and t is the time in years.

This formula is based on the concept that the percent of the whole equals the *part*, that is, percent of the principle amount is the amount of interest. You then multiply this amount by the number of years involved. It is a good idea to have this formula memorized when working with interest percentage rates. For example, to find the simple interest earned on $1,700 at 8% interest for 3 years, set up the formula $I = prt$. $I = (1,700)(0.08)(3)$, or interest = $408.00. Note that when you use the simple interest formula, you change the percent to a decimal equivalent.

If Keisha borrows $450 at an interest rate of 17% for 18 months, how much will she have paid in simple interest at the end of the 18 months?

First, remember to convert the 18 months to years: 18 months = 1.5 years. Then compute the interest owed. $I = (450)(0.17)(1.5)$, or $I = \$114.75$.

Practice 6

Fill in the blanks for the formula $I = prt$.

25. In the simple interest formula, I represents the _____.

26. In the simple interest formula, p represents the _____.

27. In the simple interest formula, r represents the _____.

28. In the simple interest formula, t represents the _____.

Often with simple interest problems you want to find the total amount of money that is either earned (in the case of an investment) or owed (in the case of a loan). To find this, you add the original principle to the interest, or $Total = p + prt$. Try your hand at these types of problems below. Answers and explanations are located at the end of the chapter.

Practice Set 7

29. Elisabeth borrows $14,000 for a new car. The simple interest is 7%, and she borrows the money for 5 years. How much interest will she repay after the 5 years?

30. Michael invests $2,500.00 in a Certificate of Deposit (CD) that pays 11% simple interest for 42 months. How much will the CD be worth in total, including interest, after the 42-month maturation date?

SUMMARY

In conclusion, here are five important things about percents you should take from this chapter:

- Percents are special ratios that compare a number to 100
- Percents are a part-whole relationship where the part is a percentage of the whole.
- For percent increase and decrease, the part is the amount of change that occurred and the whole is the original amount
- Sales problems are percent decrease problems. Tax is a percent increase problem.
- The formula for finding simple interest is $I = prt$, where I is the interest, p is the principle, r is the percentage rate (written as a decimal), and t is the time in years.

Practice Answers and Explanations

1. B

2. A

$$\frac{30}{100} = \frac{3}{10}$$

3. E

4. D

$$\frac{.8}{100} = \frac{8}{1,000}$$

5. F

6. **C**

7. **65**

Multiply 0.26 by 250 to get 65.

8. **85%**

Set up the equation $629 = 740x$. Divide both sides by 740 to get $x = 0.85$, which is 85%.

9. **45**

Set up the proportion $\dfrac{11.25}{x} = \dfrac{25}{100}$. Cross multiply to get $25x = 1125$. Divide both sides by 25 to get 45.

10. **30**

Multiply 2.00 by 15 to get 30.

11. **$4.80**

The tip amount is 15% of the total meal cost. The key word "of" means to multiply. $0.15 \times \$32.00 = \$4.80$.

12. **$3,960.00**

The commission earned is 3% of the sale price. The key word "of" means to multiply. $0.03 \times \$132,000.00 = \$3,960.00$.

13. **30,000**

Since 60% of the total commuters use public transportation, multiply: $0.60 \times 50,000 = 30,000$.

14. **37.5%**

Use a proportion, $\dfrac{\text{part}}{\text{whole}} = \dfrac{\text{percent}}{100}$. The part is the sum of the number of people who wear size 9 or size 10, or $20 + 10 = 30$. The whole is the total number of people: $10 + 15 + 25 + 20 + 10 = 80$. Fill in the missing amounts: $\dfrac{30}{80} = \dfrac{x}{100}$. Cross multiply to get $3,000 = 80x$. Divide each side by 80 to get 37.5 percent.

15. 56%

Percent increase is found by using the proportion

$\dfrac{\text{change}}{\text{original amount}} = \dfrac{\text{percent}}{100}$. The change is 14 (39-25), and the

original number is 25. By substitution, $\dfrac{14}{25} = \dfrac{x}{100}$. Cross multiply to get

1,400 = 25x. Divide both sides by 25 to get 56 percent.

16. 5.6%

The change is 5 (90-85), and the original number is 90. By

substitution, $\dfrac{5}{90} = \dfrac{x}{100}$. Cross multiply to get 500 = 90x. Divide both

sides by 90 to get 5.6 percent, rounded to the nearest tenth.

17. 16.7%

The change is 4 (24-20), and the original number is 24. By

substitution, $\dfrac{4}{24} = \dfrac{x}{100}$. Cross multiply to get 400 = 24x. Divide both

sides by 24 to get 16.7 percent, rounded to the nearest tenth.

18. 60%

The change is 75 (200-125), and the original number is 125.

By substitution, $\dfrac{75}{125} = \dfrac{x}{100}$. Cross multiply to get 7500 = 125x.

Divide both sides by 125 to get 60 percent.

19. 20%

The change in stock price is $15.00 - $12.50 = $2.50. The original

price is $12.50. By substitution, $\dfrac{2.50}{12.50} = \dfrac{x}{100}$. Cross multiply to get

250 = 12.5x. Divide both sides by 12.5 to get 20 percent.

20. 24%

The change in the number of teams is 25 - 19 = 6. The original

number is 25. By substitution, $\dfrac{6}{25} = \dfrac{x}{100}$. Cross multiply to get

600 = 25x. Divide both sides by 25 to get 24 percent.

21. $20.23

The sales tax is 7% of the purchase price of $289.00. Multiply to find the sales tax. 289.00 × 0.07 = $20.23.

22. $112.00

Since the coat is 30% off, Lara will pay 100% - 30% = 70% of the original price. Multiply 0.70 by the price of $160.00. 160.00 × 0.70 = $112.00.

23. $273.00

The discount is 25% of the $350.00 price: 0.25 × 350.00 = $87.50. The sale price is thus 350.00 - 87.50 = $262.50. The tax is 4% of this price, so multiply: 0.04 × 262.50 = $10.50. Add this to the sale price to get the total purchase price: 262.50 + 10.50 = $273.00.

24. $4,312.50

Use the fact that Josh paid 100% - 20% = 80% of the original price for the piano. Now, use a variable, x, to represent the original price. Set up the equation which shows that 80% of the original is $3,450.00: $0.80x = 3,450.00$. Divide both sides by 0.80 to get $4,312.50.

25. interest (earned or owed).

26. principle amount (invested or borrowed).

27. percentage **rate**

28. time, in years.

29. $4,900.00

This problem asks for the interest that will be paid. Use the formula $I = prt$. Interest = 14,000.00 × 0.07 × 5 = $4,900.00.

30. $3,462.50

This problem asks for the total worth of the investment, including the interest. Use the formula $Total = p + prt$. Since the time period is 42 months, first convert this to years. 42 ÷ 12 = 3.5 years. Total = 2,500.00 + (2,500.00 × 0.11 × 3.5) = $2,500.00 + $962.50 = $3,462.50.

CHAPTER 5 TEST

In this chapter, you have reviewed the key concepts of percents and have tried various types of problems. Use the chapter material, including the practice questions throughout, to assist you in solving these problems. The answer explanations that follow will provide additional help.

1. Which of the following is equivalent to 58%?

   (A) $\dfrac{5}{8}$ (B) $\dfrac{29}{50}$ (C) $\dfrac{14}{25}$ (D) $\dfrac{8}{5}$ (E) none of the above

2. Change $\dfrac{3}{8}$ to a percent.

   (A) 38%   (B) 3.8%   (C) 37.5%   (D) 0.375%   (E) 375%

3. Change 12.5% to a decimal.

   (A) 125   (B) 12.5   (C) $\dfrac{1}{8}$ (D) $\dfrac{12}{5}$ (E) 0.125

4. The number 1.2 is equivalent to what percent?
   (A) 120%   (B) 1.2%   (C) 12%   (D) 0.012%   (E) 20%

5. Which of the following sets is in increasing order?

   (A) $\{2.5, 25\%, \dfrac{3}{4}, 85\%\}$ (B) $\{25\%, \dfrac{3}{4}, 85\%, 2.5\}$

   (C) $\{25\%, 2.5, \dfrac{3}{4}, 85\%\}$ (D) $\{25\%, 85\%, \dfrac{3}{4}, 2.5\}$

   (E) none of the above

6. A county census determined that 41% of the population was 30 years old or younger. If there are 230,000 people in the county, how many are 30 years old or younger?

   (A) 560,976   (B) 94,300   (C) 230,041

   (D) 89,700   (E) 589,744

7. What is 62% of 434?

   (A) 269.08   (B) 700   (C) 2,690.8   (D) 70   (E) 2.6908

8. What percent of 500 is 90?

   (A) 5.56%   (B) 0.18%   (C) 55.6%

   (D) 18%   (E) 90%

9. What is 82% of 112?

   (A) 82   (B) 91.84   (C) 30   (D) 20.16   (E) 0.7321

10. Forty-five percent of the college students at the local university are not registered to vote. If the student population is 8,500 students, how many students are not registered to vote?

    (A) 4,000   (B) 450   (C) 4,500   (D) 4,675   (E) 3,825

11. Of the 200 cars in the used car sales lot, 80 of them are white. What percent of the cars are white?

    (A) 40%   (B) 80%   (C) 120%   (D) 0.40%   (E) 12%

12. The number of people enrolled at the health spa rose from 240 to 310. What is the percent increase, to the nearest percent?

    (A) 70%   (B) 77%   (C) 29%   (D) 23%   (E) 31%

13. What is the percent decrease from 72 to 64, to the nearest percent?

    (A) 11% (B) 12% (C) 8% (D) 89% (E) 17%

14. Using the graph for stock prices below, what is the percent increase in stock share price from 2002 to 2003?

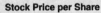

Stock Price per Share

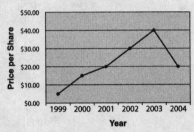

    (A) 10% (B) 25% (C) 33.3% (D) 75% (E) 50%

15. What is the percent increase in sales, to the nearest percent, if last years sales were 12,000 and this years sales are 25,000?

    (A) 48% (B) 13% (C) 108% (D) 10.8% (E) 52%

16. A tent has a sale price of $450.00. This price reflects a 25% discount. What was the original price of the tent?

    (A) $475.00 (B) $425.00 (C) $1,800.00
    (D) $600.00 (E) $180.00

17. What is the sales tax on a $49.00 cell phone purchase, if the sales tax percentage is 7%?

    (A) $56.00 (B) $7.00 (C) $34.30
    (D) $5.60 (E) $3.43

18. All novels at the convenience store are discounted 20%. There is a 4% sales tax on books. What is the total purchase price of a book listed at $14.95, including tax?

    (A) $12.44   (B) $11.96   (C) $3.59

    (D) $12.56   (E) $11.36

19. Given a tax amount of $7.80 on a snowboard, priced at $130.00, what is the sales tax percentage?

    (A) 7.8%   (B) 0.06%   (C) 6%   (D) .078%   (E) 60%

20. Using the bar graph below, approximately what percentage of the day is spent sleeping?

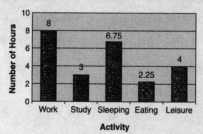

Time Allocation per Average Day

    (A) 6.75%   (B) 28%   (C) 33.3%

    (D) 12.5%   (E) 8%

21. Robert borrows $75,000 at 5.5% simple interest for his new home mortgage. If the mortgage is held for 30 years, how much total will he spend on the loan and interest?

    (A) $198,750.00   (B) $123,750.00   (C) $12,375.00

    (D) $87,375.00   (E) $165.00

22. Rhonda invests $3,500.00 in a savings account that pays 7% simple interest. How much will she have in her account after 30 months, if no other deposits or withdrawals are made?

    (A) $612.50   (B) $7,350.00   (C) $3,710.00

    (D) $4,112.50   (E) $48,125.00

Answers and Explanations

1. B

Change 58% to a fraction, and then simplify by dividing by the common factor of 2: $\frac{58}{100} \div \frac{2}{2} = \frac{29}{50}$.

2. C

Set up a proportion: $\frac{3}{8} = \frac{x}{100}$, where x is the percent. Cross multiply, to get $8x = 300$. Divide both sides by 8 to get $x = 37.5\%$. If your answer choice was (A), you thought that $\frac{3}{8}$, meant 38%, which is incorrect. 38% is equal to $\frac{38}{100} = \frac{19}{50}$. In choice (B), 3.8% is equal to 0.038, or $\frac{38}{1,000} = \frac{19}{500}$. Choice (D) is the decimal equivalent of $\frac{3}{8}$: $3 \div 8 = 0.375$.

The percent is found by multiplying this decimal by 100. If you chose (E), you may have converted the fraction to a decimal, 0.375, and then mistakenly interpreted this to be 375%.

3. E

To change a percent to a decimal, just move the decimal point 2 places to the left to get 0.125.

4. A

To change a decimal to a percent, just move the decimal point 2 places to the right, adding a trailing 0, to get 120%.

5. B

Change each of the numerical values to a decimal to compare them.
$25\% = 0.25$, $\frac{3}{4} = .75$ and $85\% = 0.85$. It is easier now to see that
$0.25 < 0.75 < 0.85 < 2.5$.

6. B

The problem states that 41% of the population is 30 years or younger.
The key word "of" means to multiply: $0.41 \times 230,000 = 94,300$.

7. A

Change 62% to a decimal and multiply: $0.62 \times 434 = 269.08$.

8. D

Set up an equation, stating that part = percent × whole. Use the
variable x to represent the percent, written as a decimal. $90 = 500x$.
Divide both sides by 500 to get $x = 0.18$, or 18%.

9. B

Change 82% to a decimal and multiply: $0.82 \times 112 = 91.84$.

10. E

This is an application of percent based on the part-whole relationship.
Change 45% to a decimal and then multiply by the whole student
body of 8,500. $0.45 \times 8,500 = 3,825$.

11. A

Set up a proportion using $\frac{\text{part}}{\text{whole}} = \frac{\%}{100}$, where 80 is the part and

200 is the whole. Use the variable x for the percent. $\frac{80}{200} = \frac{x}{100}$.
Cross multiply to get $200x = 8,000$. Divide both sides by 200 to get
$x = 40\%$.

12. C

Use the proportion: $\frac{\text{change}}{\text{original amount}} = \frac{\text{percent}}{100}$. The change in

enrollment is $310 - 240 = 70$. The original amount is 240. Let x

represent the percent increase: $\frac{70}{240} = \frac{x}{100}$. Cross multiply to get

$240x = 7,000$. Divide both sides by 240 to get $x = 29.17$, or 29%.

13. A

Use the proportion: $\dfrac{\text{change}}{\text{original amount}} = \dfrac{\text{percent}}{100}$. The change is

$72 - 64 = 8$. The original amount is 72. Let x represent the percent

decrease: $\dfrac{8}{72} = \dfrac{x}{100}$. Cross multiply to get $72x = 800$. Divide both

sides by 72 to get $x = 11.11$, or 11%.

14. C

Read the values for the stock price for 2002 and 2003. These values
are $30.00 and $40.00, respectively. Use the proportion:

$\dfrac{\text{change}}{\text{original amount}} = \dfrac{\text{percent}}{100}$. The change is $40 - 30 = 10$. The original

amount is 30. Let x represent the percent increase: $\dfrac{10}{30} = \dfrac{x}{100}$. Cross

multiply to get $30x = 1,000$. Divide both sides by 30 to get $x = 33.3\%$.

15. C

Use the proportion: $\dfrac{\text{change}}{\text{original amount}} = \dfrac{\text{percent}}{100}$. The change is 25,000

$- 12,000 = 13,000$. The original amount is 12,000. Let x represent the

percent increase: $\dfrac{13,000}{12,000} = \dfrac{x}{100}$. Cross multiply to get $12,000x = $

$1,300,000$. Divide both sides by 12,000 to get $x = 108.33$, or 108%.

16. D

If the tent is on sale for 25% off, then the sale price is 100% - 25% =
75% of the original price. Use the variable x to represent the original
price of the tent, remembering that the part (the sale price), is a
percent (75%), of the whole: $450.00 = 0.75x$. Divide both sides by
0.75 to get $x = \$600.00$.

17. E

If the sales tax percentage is 7%, then the sales tax is 7%, or 0.07,
of the purchase price. The key word "of" means to multiply:
$0.07 \times 49.00 = \$3.43$.

18. A

The novels are discounted 20%, so the buyer will pay 100% - 20% = 80% of the listed price. $0.80 \times 14.95 = \$11.96$. Because sales tax is added to the price, the buyer will also now pay 100% + 4% = 104% of the sale price: $1.04 \times 11.96 = \$12.44$.

19. C

The sales tax is a part of the purchase price. Use the proportion $\frac{change}{original\ amount} = \frac{\%}{100}$. The part is the $7.80 tax amount, the whole is the $130.00 purchase price and x will represent the sales tax percentage. $\frac{7.80}{130.00} = \frac{x}{100}$. Cross multiply to get $130x = 780$. Divide both sides by 130 to get $x = 6\%$.

20. B

Use the proportion $\frac{part}{whole} = \frac{\%}{100}$. The part is the hours spent sleeping, or 6.75, the whole is a 24 hour day and x will represent the percentage. $\frac{6.75}{24} = \frac{x}{100}$. Cross multiply to get $24x = 675$. Divide both sides by 24 to get $x = 28\%$.

21. A

For this problem, you are asked to find the total he will spend on the loan and the interest. Use the simple interest formula: $Total = P + PRT$, where P is the principle borrowed (75,000), R is the percentage rate (5.5%, or 0.055), and T is the time (30 years). Total = 75,000 + $75,000 \times 0.055 \times 30$, or Total = 75,000 + 123,750 = \$198,750.00.

22. D

For this problem, you are asked to find the total in the account which includes the deposit plus the interest. Use the simple interest formula: $Total = P + PRT$, where P is the principle deposited (3,500), R is the percentage rate (7%, or 0.07), and T is the time (30 months or 2.5 years). Total = 3,500 + $3,500 \times 0.07 \times 2.5$, or Total = 3,500 + 612.50 = \$4,112.50

CHAPTER 6

Statistics and Probability

BUILDING BLOCK QUIZ

The lessons in this chapter will clarify and extend your knowledge of statistics and probability. Take the following Building Block Quiz to assess your understanding of the topics studied in the last chapter, and to explore the new concepts to be covered in this chapter.

1. What percent of 140 is 35?

    (A) 4% (B) 15% (C) 25% (D) 35% (E) 40%

2. The percent 37.5% can also be expressed as which of the following?

    (A) $\dfrac{3}{8}$ (B) $\dfrac{1}{8}$ (C) $3\dfrac{1}{4}$ (D) $37\dfrac{1}{2}$ (E) $\dfrac{7}{16}$

3. What is the total amount of simple interest earned on $400 invested at 5% per year for 3 years?

    (A) $20 (B) $60 (C) $600 (D) $460 (E) $6,000

4. The height in inches of 7 students in a class are 63, 66, 69, 64, 59, 63, 72. What is the median height in inches?

    (A) 59 (B) 63 (C) 64 (D) 66 (E) 72

5. The temperature in degrees Fahrenheit on four consecutive days was 4, –2, –8, and 6. What is the mean temperature over the four days?

   (A) –8   (B) –2   (C) 0   (D) 6   (E) none of these

6. Given the following set of data, which of the following statements is NOT true?

   {2, 5, 6, 8, 10, 12, 20}

   (A) The mode is 8.
   (B) The range is greater than the median.
   (C) The median is 8.
   (D) The mean is 9.
   (E) The mean is less than range.

7. The following graph shows the weekly pizza sales at a pizzeria for four different kinds of pizza. If a total of 200 pizzas were sold, how many more pepperoni pizzas were sold than mushroom pizzas?

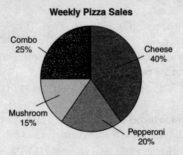

Weekly Pizza Sales

Combo 25%
Cheese 40%
Mushroom 15%
Pepperoni 20%

   (A) 100   (B) 70   (C) 40   (D) 30   (E) 10

8. When spinning the spinner below, what is the probability of getting an odd number?

(A) $\frac{1}{8}$ (B) $\frac{3}{8}$ (C) $\frac{1}{4}$ (D) $\frac{1}{2}$ (E) $\frac{5}{8}$

9. When selecting a card from a standard deck, what is the probability of selecting a king or a red card?

(A) $\frac{26}{52}$ (B) $\frac{28}{52}$ (C) $\frac{30}{52}$ (D) $\frac{32}{52}$ (E) $\frac{34}{52}$

10. What is the total number of five-digit zip codes that can be formed if the digits can repeat and the first number in the code must be a 1 or a 3?

(A) 120 (B) 2,000 (C) 20,000
(D) 30,240 (E) 100,000

Answers and Explanations

1. C

Set up the proportion $\frac{35}{140} = \frac{x}{100}$. Use cross-multiplication to get the equation $3500 = 140x$. Divide each side of the equation by 140 to get $x = 25$. 35 is 25% of 140.

2. A

Because 37.5% is a percent, you can write it as the fraction $\frac{37.5}{100}$.

Make the numerator a whole number by multiplying both numerator and denominator by 10. $\frac{37.5}{100} = \frac{375}{1000}$. Divide both numerator and denominator by the greatest common factor of 125 to simplify the fraction to $\frac{3}{8}$. Another strategy to solve this problem is to convert each answer choice to decimal form by dividing the numerators by the denominators. The fraction equal to 0.375, which is $\frac{3}{8}$, will be equivalent to 37.5%.

3. B

This problem asks you to calculate simple interest. The formula for simple interest is $I = prt$, where I = interest, p = principle invested, r = yearly interest rate in decimal form, and t = time in years. Substitute the known values into the formula: $I = (400)(0.05)(3)$. Multiply these values together to get the result of $60.

4. C

This question tests the concept of median. The median height will be the middle number in the list when the numbers are arranged in order from smallest to largest. The set in this order becomes: 59, 63, 63, 64, 66, 69, 72. Since there are 7 numbers in the list, the fourth number, 64, is the median. Choice (A) is the lowest number in the set, not the middle number. Choice (B) is the mode, or most frequent number in the set, not the middle number. Choices (D) and (E) are too large to be the middle number in the set.

5. C

This problem addresses the concept of mean. To find the mean, find the sum of the list of numbers and divide the sum by the total number of data values in the list. The sum is $4 + -2 + -8 + 6 = 0$. Since there are four numbers in the list, divide the sum by 4: $0 \div 4 = 0$.

6. A

This questions tests your knowledge of various statistical measures. To solve this question, it may be helpful to find a few of the measures mentioned in the answer choices. Starting with choice (A): The mode is the number that appears most often, and in this list there is no mode, making (A) false and the correct answer. Since the numbers are already listed in ascending order, the median is 8, making choice (C) true. The range is the smallest value subtracted from the largest value: $20 - 2 = 18$. Since 18 is greater than 8, the range is greater than the median and answer choice (B) is true. The sum of the list of numbers is 63; $63 \div 7 = 9$, so the mean is 9, making answer choice (D) true. Since 9 is less than 18, then answer choice (E) is also true because the mean is less than the range.

7.　E

The problem assesses your ability to work with circle graphs. To find the number of pepperoni pizzas, find 20% of 200. Set up the proportion $\frac{20}{100} = \frac{x}{200}$. Cross-multiply to get $4,000 = 100x$; divide each side of this equation by 100 to get 40. To find the number of mushroom pizzas, find 15% of 200 using the same procedure. Set up the proportion $\frac{15}{100} = \frac{x}{200}$, cross-multiply to get $3,000 = 100x$, then divide to get $x = 30$. Since $40 - 30 = 10$, there were 10 more pepperoni pizzas sold than mushroom pizzas. Another approach to this question is to subtract the percents first; $20 - 15 = 5\%$. Then find 5% of 200. This number will be the difference in the number of pizzas. Set up the proportion $\frac{5}{100} = \frac{x}{200}$, cross-multiply to get $1,000 = 100x$, and divide to get $x = 10$.

8.　D

This question tests the concept of simple probability. Probability is $\frac{\text{the number of ways event can occur}}{\text{the total number of possible outcomes}}$. Since there are 8 equal sections on the spinner, and 4 contain an odd number (1, 3, 5, or 7), then the probability is $\frac{4}{8}$, which simplifies to $\frac{1}{2}$.

9.　B

This question deals with the concept of compound probability.

The probability of a compound event is $P(A \text{ or } B) = P(A) + P(B) -$

$P(A \text{ and } B)$. The probability of selecting a king from a standard deck of

52 cards is $P(\text{king}) = \dfrac{4}{52}$. The probability of selecting a red card is

$P(\text{red card}) = \dfrac{26}{52}$. Since there are 2 kings that are also red, the $P(\text{king}$

and a red card$) = \dfrac{2}{52}$. The equation becomes $P(\text{king or red card}) =$

$\dfrac{4}{52} + \dfrac{26}{52} - \dfrac{2}{52} = \dfrac{28}{52}$.

10. C

This question tests your knowledge of counting problems. Take the number of possibilities of each digit in the code and multiply them together. Since the first digit must be a 1 or a 3, there are only 2 choices for the first digit. Since the numbers can repeat, each of the other digits can be 0 through 9, which is 10 different numbers. Therefore, the total number of zip codes can be represented by $2 \times 10 \times 10 \times 10 \times 10 = 20,000$.

STATISTICS AND PROBABILITY

This chapter will review the concepts of probability and statistics. You will learn about new concepts including measures of central tendency such as mean, median, mode, and range, different types and uses of statistical graphs, probability, and counting problems.

Measures of Central Tendency

Measures of central tendency are values that are examined from a set of data in order to make predictions and draw conclusions about that set of data as a whole. Four of the most common measures that we use are mean, median, mode, and range.

Mean

> ### MATH SPEAK
> The **mean** of a set of numbers is the sum of the numbers in the set divided by the total number of values in the set.

For example, in order to find the mean of the set {8, 9, 22, 14, 12}, first find the sum by adding the numbers: $8 + 9 + 22 + 14 + 12 = 65$. Since there are 5 numbers in the set, the mean can be found by dividing the sum of 65 by 5. The mean is $65 \div 5 = 13$.

Another variation of a problem involving mean is a question where one or more values from the data set are unknown, but the mean is identified. Take the following example:

Michelle received an 88, 84, 79, and 93 on her first 4 science exams. What does she need on her fifth exam to have an exam average (arithmetic mean) of 88?

There are a few different ways to approach this problem. One is to find the total number of points Michelle has earned so far, and to compare that number with the amount of points she needs to have a mean of 88. The sum of her first 4 tests is $88 + 84 + 79 + 93 = 344$ points. The number of points she needs to have an 88 average over 5 exams is $88 \times 5 = 440$. Since the number of points she needs is 440 and the number of points she has earned so far is 344, subtract those amounts to find the points she needs on the 5th test: $440 - 344 = 96$.

An algebraic approach to this problem is to use the proportion

$$\frac{\text{sum of all values}}{\text{number of values}} = \text{average (arithmetic mean)}$$

If you call the fifth test the unknown score x, this proportion is written as $\dfrac{88 + 84 + 79 + 93 + x}{5}$. Simplifying by adding the known values makes the equation $\dfrac{344 + x}{5} = 88$. Since 88 can also be represented as $\frac{88}{1}$, cross-multiply the values to get the equation $344 + x = 440$.

Subtract 344 from each side of the equation to get $x = 440 - 344 = 96$. Either method results in a fifth exam grade of 96.

> ### FLASHBACK
> For help with **cross-multiplying**, refer to chapter 4, Ratio and Proportion.

Median

> ### MATH SPEAK
> The **median** of a set of data is the middle number when the values are listed in order from smallest to largest.

In lists containing an odd number of values, one number will be located directly in the middle of the list. If there is an even amount of numbers in the list, two numbers will share the middle. In this case, find the average of those two values to find the median of the entire list.

Find the median of the following set of numbers
$\{20, 65, 34, 21, 55, 89, 38, 41, 76\}$

First, list the numbers in order from smallest to greatest.

20, 21, 34, 38, 41, 55, 65, 76, 89

Since there are nine numbers in the list, the fifth number, 41, is in the middle. The number 41 is the median of the data set.

Mode

> **MATH SPEAK**
> The **mode** of a set of data is the number that occurs the most in a data set.

Sometimes we encounter a list where there is not a number that occurs more than any other in the list; in this case we say there is *no mode*. There is also the case where there may be more than one value that occurs as much as another number. If two numbers occur the same number of times and they occur more than any other number in the list, the set is considered to be *bimodal*, meaning there are two modes. It is also possible to have more than two modes in a set of data.

Find the mode in each data set:

8, 9, 10, 4, 5, 8, 8	The mode is 8.
2, 20, 18, 66, 3, 14, 23	There is no mode for this set.
−1, 9, 10, −1, 5, 77, 10	Both −1 and 10 appear the most. This set is bimodal.

Range

> **MATH SPEAK**
> The **range** of a set of numbers is the difference between the largest value in the list and the smallest value in the list.

Take the following data set: 34, 54, 22, 84, 90, 55, 60, and 23. The range of this set would be $90 - 22 = 68$, since 90 is the largest value in the set and 22 is the smallest.

Practice 1

Try the following matching exercises in order to polish your skills finding mean, median, mode, and range. Answers and explanation are located at the end of the chapter.

1. The mean of the set   A. 27
   {9, 7, 8, 14, 22, 12}

2. The mode of the set   B. 13
   {22, 33, 91, 45, 33, 76, 87}

3. The range of the set   C. 12
   {101, 105, 112, 123, 98, 96}

4. The sum of the median and   D. 36
   the range of the set {1, 4, 9, 8, 5}

5. The median of the set   E. 33
   {24, 53, 32, 39, 33, 61}

Statistical Graphs

There are many types of statistical graphs used to display data. Some of the most common types are the bar graph, the line graph, the circle graph, the scatter plot, and the histogram.

Bar graphs are used to compare data. They can be drawn using horizontal bars or vertical bars. The graph below compares the price of a popular video game from four different stores.

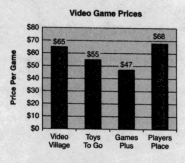

Using the given graph, answer the following question:

How much more does a video game cost at Video Village than at Games Plus?

By reading the above graph, the price of the video at Video Village is $65 and the price at Games Plus is $47. Since the question asks for the difference in price, subtract the 2 values: $65 − 47 = $18.

A **line graph** is used to show a trend that occurs over time. Take, for example, the following graph that shows stock prices for a certain company over one month.

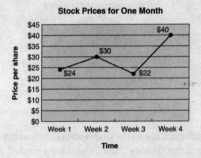

Using the information in the graph above, answer the following question:

Between which two weeks did the price of the stock almost double?

You are looking for an increase in price. The price week 3 was $22 and the price week 4 was $40. Since $22 times 2 is equal to $44 which is close to $40, then between these two weeks the price of the stock almost doubles.

A **circle graph** is used to show parts of whole. Percents are often used within a circle graph; always make sure that any percents used in a circle graph always add to 100%.

> ### FLASHBACK
>
> To review the steps for calculating with **percent**, refer to chapter 5, Percent.

The following graph shows the breakdown of the participation in extracurricular activities at a middle school. Each student can be a member of only one club.

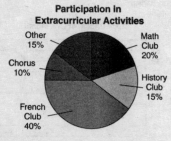

Participation in Extracurricular Activities

Other 15%

Math Club 20%

Chorus 10%

History Club 15%

French Club 40%

From the information in the graph above, how many students participate in the History Club if there are a total of 420 students who take part in an extracurricular activity?

Since 15% of the students participate in History Club, find 15% of 420.

Use the proportion $\frac{x}{420} = \frac{15}{100}$. Cross-multiply to get $100x = 6,300$.

Divide each side of the equation by 100 to get $x = 63$. Sixty-three students participate in History Club.

> ### REMEMBER THIS!
>
> A **bar graph** is used to compare data.
>
> A **line graph** is used to show trends over time.
>
> A **circle graph** is used to show parts of a whole; the percents in a circle graph must add to 100%.

Practice 2

In this practice set, match the information with the best choice of graph. Answers and explanations are found at the end of the chapter.

6. The Wood family's monthly budget

Rent	40%
Food	35%
Utilities	15%
Other	10%

A. bar graph

7. The high temperature each day five days

Monday	61°
Tuesday	59°
Wednesday	60°
Thursday	54°
Friday	72°

B. line graph

8. The number of employees at four different companies

Company A	450
Company B	300
Company C	490
Company C	110

C. circle graph

Scatter Plots

Much like a line graph, a scatter plot is also used to show trends in data. In this type of graph, however, the data points are not connected. Instead, the general trend of the data points is examined. The more the data in the graph appears to form a straight line, the stronger the relationship is between the data in the graph.

The following is an example of a scatter plot that shows the relationship between the number of assignments completed and the class average from a certain academic class.

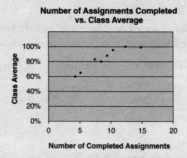

Number of Assignments Completed vs. Class Average

Specific information about the data can be determined from the graph. For example, the greatest number of assignments completed is 15, and the person who completed 15 assignments has approximately a 98% class average.

In a scatter plot, there is usually one of three conclusions that can be made based on the graph. The first is a positive trend, or positive correlation: when the *x*-values increase there is also an increase in the *y*-values. As you can see from the graph above, the relationship between the number of assignments completed and the class average in this class is positive. The more assignments completed, the higher the class average. The less assignments completed, the lower the class average.

The second type of trend is a negative trend, or negative correlation: when there is a decrease in the *y*-values as the *x*-values increase. An example of a negative correlation might be the cost of a ticket for a show and the number of tickets sold. As the price increases, the number of tickets purchased may decrease.

A third possibility is that there is no correlation, or relationship, in the data graphed. The figure below shows what each type of correlation looks like on a graph.

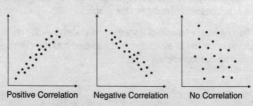

Positive Correlation   Negative Correlation   No Correlation

Stem and Leaf Plots

Before constructing certain types of graphs, it is very useful to organize your data in a chart or table for easy reference. One way to organize a list of numbers is a stem and leaf plot. In the stem and leaf plot, there are typically two columns. The first column, known as the stem, often represents the first digit(s) in the number and the second column, known as the leaf, represents the last digit of the number. For example, if placing the number 29 in a stem and leaf plot, the 2 would appear in the stem and the 9 would appear in the leaf.

stem	leaf
2	9

When dealing with more than two digits, such as three-digit numbers, the first two digits would form the stem and the number in the ones place would be the leaf.

In a stem and leaf plot, each of the stems is listed in ascending order so that you can easily read and interpret the information. The figure below shows a stem and leaf plot containing 10 data values.

stem	leaf
2	8, 9
3	4
4	1, 2, 9
5	0, 1, 1, 4

The numbers in the plot are 28, 29, 34, 41, 42, 49, 50, 51, 51, and 54. Notice a few things:

a. You can tell the total number of values in the list by counting the leaves. Here there are a total of 10 leaves.

b. You can tell if a number repeats because there will be a leaf with a number or numbers that repeat. The stem of 5 has a leaf of 1 that repeats, so the number 51 is in the data set twice.

c. You can also figure out measures such as mode, median, and range quickly since the numbers are in order. The mode is 51, the median is 45.5 and the range is 54 − 28 = 26.

Histograms

One type of graph where stem and leaf plots are particularly helpful is the histogram. Histograms appear to be much like bar graphs except for two major differences: the bars in the histogram represent number values or intervals of values, and the bars are touching each other.

Assume that the data in the above stem and leaf plot represented the attendance at various performances of a play. The following is an example of a histogram made from the data in the above stem and leaf plot. Notice that since the data is already set up in intervals of 10, constructing the histogram is quick and easy.

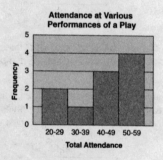

Attendance at Various Performances of a Play

Using the graph above, answer the following question:

How many performances had an attendance between 40 and 60 people?

To find this solution, add the total frequency of the bar for 40 – 49 people with the frequency of the bar for 50 – 59 people. $3 + 4 = 7$ performances.

Being able to read and interpret different types of statistical graphs is a skill that is not only useful on high-stakes exams, but also in daily life. Try the following practice questions to test your skills on analyzing graphs.

Practice 3

Write the result on the answer blank that best answers each question. Answers and explanations are located at the end of the chapter.

9. Based on the scatter plot below, a person with 5 years of experience makes how much per hour? _____

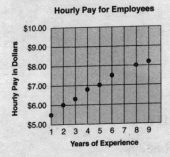

Hourly Pay for Employees

10. According to the stem and leaf plot below, what is the range of the set of data? _____

stem	leaf
1	0, 5
2	6
3	7, 8, 9
4	1, 1, 2, 3
5	5, 8
6	8, 8, 9

11. According to the histogram below, what interval is the mode of the set of grades? _____

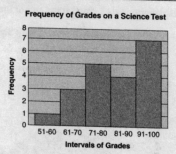

Frequency of Grades on a Science Test

Intervals of Grades

PROBABILITY

Simple Probability

You have probably experienced a time when you have made a guess on the likelihood of an event, such as the chance it might rain, or landing on heads when flipping a coin. Theoretical probability is the principle used when calculating the chance that these and other events could happen.

> **MATH SPEAK**
>
> The **probability** of an event (E) is defined as
>
> $$P(E) = \frac{\text{the number of ways event E can occur}}{\text{the total number of possible outcomes}}$$

The *outcome* set, also known as the sample space, is the list of all possible outcomes. For example, the outcome set when flipping a coin is {Heads, Tails} because these are the only possible outcomes. To calculate the probability that you will get tails when flipping a coin, $P(\text{tails}) = \frac{1}{2}$ since there is 1 way to get tails out of 2 possible outcomes.

> ## REMEMBER THIS!
> The probability of an event that can't happen, or an impossible event, is 0.
>
> The probabiltiy of an event that is certain to happen is 1.
>
> All other simple probabilities will be between 0 and 1.

Compound Probability

Compound probability occurs when a problem asks for the chance of more than one outcome to occur. The key word used in most compound probability questions is "or."

> ## REMEMBER THIS!
> The formula for compound probability is
>
> $$P(A \text{ or } B) = P(A) + P(B) - P(A \text{ and } B).$$
>
> **Mutually exclusive** events are situations that cannot occur at the same instance, so P(A and B) from the formula above will be 0.

Two examples of compound probability are finding the probability of getting a 3 or a 6 when rolling a die, or finding the probability of selecting a red card or a 5 from a standard deck of 52 cards. In the first example, the probability of rolling a 3 is $\frac{1}{6}$, the probability of rolling 6 is $\frac{1}{6}$, and the probability of rolling both a 3 *and* a 6 is $\frac{0}{6}$. Thus, the probability is

$$P(3 \text{ or } 6) = \frac{1}{6} + \frac{1}{6} - \frac{0}{6} = \frac{2}{6} = \frac{1}{3}.$$

Note that since the two events were mutually exclusive, the probability that *both* would happen was $\frac{0}{6}$ or 0.

In the second example, the probability of selecting a red card is $\frac{26}{52}$, the probability of selecting a 5 is $\frac{4}{52}$, and the probability of selecting a red card that is *also* a 5 is $\frac{2}{52}$. Thus, the probability is

$$P(\text{red card or a 5}) = \frac{26}{52} + \frac{4}{52} - \frac{2}{52} = \frac{28}{52} = \frac{7}{13}.$$

In this example, these two events could occur at the same time. In other words, there was a possibility that a card chosen could be a red 2 (such as the 5 of hearts or 5 of diamonds). Since each of those cards fits both categories of being red and being a 5, they can not be counted twice. Therefore, the chance that 1 of those 2 cards was selected ($P(A \text{ and } B) = \frac{2}{52}$) was subtracted out.

Practice 4

Fill in the blank with the solution that best answers the question. Answers and explanations are located at the end of the chapter.

12. When rolling one die, the probability of rolling a 6 is

    _____.

13. When spinning the spinner below, the probability of getting section B is _____.

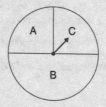

14. When selecting a card from a standard deck, the probability of choosing a 2 or a face card is _____.

15. When rolling one die, the probability of rolling a 3 or an odd number is _____.

16. If a bag contains only 5 red, 6 purple and 2 balls, then P(pink ball) is _____.

Independent and Dependent Events

More complex situations occur in probability problems when more than one event is taking place at the same time, or when one is taking place right after another. In these circumstances, the probabilities of each event are multiplied together to get the final likelihood of occurance.

REMEMBER THIS!

The key word used often in **independent and dependent events** is "and," which indicates that the probabilities should be multiplied together.

Independent Events

Independent events are two or more events that will occur without the outcome of one affecting the outcome of any other. The probability of the second event will be the same regardless of the outcome of the first event. An example of two independent events is rolling a die and flipping a coin. No matter what was rolled on the die, the probability of getting heads or tails on the coin will always be the same. Another example would be selecting a card from a deck, replacing the card, and then selecting the next card. In this case, the key word *replace* or the phrase *with replacement*, would indicate that the events were independent. Since the first card would be replaced, there will always be 52 cards to choose from when selecting a card. The probabilities would be the same each time.

Try this exercise with two independent events:

A bag contains 3 red marbles, 5 blue marbles, and 2 green marbles. A marble is selected at random, replaced, and then another marble selected. What is the probabililty of P(red, blue)?

This question is asking you to find the probability of pulling out a red marble first and then a blue marble second. Since the marble is selected, and then replaced, there are always the same 10 marbles to choose from in the bag. Therefore, the events are independent of each other. The probability of selecting a red marble first is $\frac{3}{10}$, and the probabililty of selecting a blue marble second is $\frac{5}{10}$. Thus, P(red, blue) $= \frac{3}{10} \times \frac{5}{10} = \frac{15}{100}$ or $\frac{3}{20}$.

Dependent Events

Dependent events are situations where the outcome of the first event *does* affect the probability of the second event. These are often events presented much like the ones mentioned above, but there is *no replacement*. For example, find the probability of selecting 2 kings from a standard deck of cards *without replacement*. The probability of selecting a king is $\frac{4}{52}$. Since the first king is not replaced, the probability of selecting a second king is only $\frac{3}{51}$. There are only 3 kings left and a total of 51 cards after the first one is selected and not put back in the deck. Thus, the probabililty for selecting 2 kings from a deck without replacement is $\frac{4}{52} \times \frac{3}{51} = \frac{12}{2652}$, or $\frac{1}{221}$.

Take another look at the exercise from above, but now with two dependent events:

A bag contains 3 red marbles, 5 blue marbles, and 2 green marbles. A marble is selected at random, not replaced, and then another marble selected. What is the probabililty of P(red, blue)?

Since the marble is selected, and then not replaced, there will only be 9 marbles left in the bag for the second draw. Therefore, the events are dependent on each other. The probability of selecting a red marble first is $\frac{3}{10}$, and the probabililty of selecting a blue marble second is $\frac{5}{9}$.

Thus, P(red, blue) = $\frac{3}{10} \times \frac{5}{9} = \frac{15}{90}$ or $\frac{1}{6}$.

Don't take chances with your skills in probability! The following practice set will provide additional practice in finding the probability of independent and dependent events.

Practice 5

Answer true or false to each of the following statements. Answers and explanations are located at the end of the chapter.

17. **T   F**   If two cards are drawn from a standard deck *with replacement*, the probability of drawing a king and then a queen is $\frac{16}{2,704}$ or $\frac{1}{169}$.

18. **T   F**   When rolling a die and flipping a coin, the probability of P(5, tails) is $\frac{1}{8}$.

19. **T   F**   If two cards are drawn from a standard deck *without replacement*, the probability of drawing a king and then a queen is $\frac{12}{2,652}$ or $\frac{1}{221}$.

20. **T   F**   A bag contains 3 red, 5 blue, and 4 white marbles. If a marble is drawn from the bag, not replaced, and then another is drawn, then P(white, blue) is $\frac{5}{33}$.

The Counting Principle

Sometimes when you are given many choices to select from, the number of possible choices can seem unlimited. However, when the number of choices in a situation is known, the actual number of possibilities can be calculated. For example, when ordering a sandwich at a deli you may have 3 different choices of condiments, 4 choices of bread, and 5 choices of meat, or main ingredient, for the sandwich. To find the total number of possible sandwich combinations, multiply the number of choices in each category together: $3 \times 4 \times 5 = 60$ different sandwiches. Multiplying the number of choices together is a concept called the **counting principle**. This strategy is also used when solving the next two types of situations presented: permutations and combinations.

Permutations

> ### MATH SPEAK
>
> A **permutation** of objects is the number of possible arrangements for that set of objects.

With permutations, it's all about order. Every time the order changes, a new permutation of the objects is formed. If 6 different books are placed on a shelf, there are 6 choices for the first spot on the shelf, 5 choices for the second, 4 for the third, 3 for the fourth, 2 for the fifth, and only 1 book left for the sixth. Thus, the total number of arrangements for those 6 books can be represented by an operation that looks like this: $6! = 6 \times 5 \times 4 \times 3 \times 2 \times 1 = 720$. This is known as 6 *factorial*.

> ### MATH SPEAK
>
> The **factorial** of a whole number is the product of that whole number and each of the natural numbers less than the number. It is written as $n! = n \times (n-1) \times (n-2) \times \ldots \times 1$.

In other words, the number of permutations of n objects taken n at a time is $_nP_n = n!$.

But sometimes not all of the objects are considered for each different arrangement. In this case, n represents the total number of objects you have to choose from and r is the number of objects actually selected to be arranged.

> ## MATH SPEAK
>
> The formula for the permutation of n objects taken r at a time is
> $$_nP_r = \frac{n!}{(n-r)!}.$$

For example, if there are 7 runners in a race, how many different orders are there for first, second and third place? In this case there are a total of 7 runners but you are only finding the number of permutations of 3 of them at a time. Using the above formula,

$$_nP_r = \frac{n!}{(n-r)!}$$

$$= {_7P_3} = \frac{7!}{(7-3)!} = \frac{7!}{4!} = \frac{7 \times 6 \times 5 \times 4 \times 3 \times 2 \times 1}{4 \times 3 \times 2 \times 1}$$

$$= \frac{7 \times 6 \times 5 \times \cancel{4} \times \cancel{3} \times \cancel{2} \times \cancel{1}}{\cancel{4} \times \cancel{3} \times \cancel{2} \times \cancel{1}} = 7 \times 6 \times 5 = 210 \text{ different ways}$$

A faster way to think of this is by using the counting principle. There are 3 different places (first, second, third) to be considered. Since there are 7 different choices for first place, then 6 different choices for second place, then 5 different choices for third place, the number of arrangements is $7 \times 6 \times 5 = 210$.

Combinations

With combinations, the order is not important. The grouping of objects changes the number of combinations that exist.

> ## MATH SPEAK
>
> A **combination** is the total number of groupings of a set of objects. The formula for the number of combinations of n objects taken r at a time is $_nC_r = \frac{n!}{r!(n-r)!}.$
>
> Note that $_nC_n$ will always be equal to 1. There is only one way to select *all* the members of a group or committee.

For example, if you were selecting 3 people from a group of 10 to form a committee, the order of the members of the committee is not important; each person, whether he is picked first, second, or third, will end up being on the committee. In this way, the key words for combinations include the words "choose" and "committee."

Therefore, if there are 10 people to choose from for a committee of 3, how many different combinations could be formed? Using the above formula,

$$_nC_r = \frac{n!}{r!(n-r)!}$$

$$= {_{10}C_3} = \frac{10!}{3!(10-3)!} = \frac{10!}{3!7!}$$

$$= \frac{10 \times 9 \times 8 \times 7 \times 6 \times 5 \times 4 \times 3 \times 2 \times 1}{3 \times 2 \times 1 \times (7 \times 6 \times 5 \times 4 \times 3 \times 2 \times 1)}$$

$$= \frac{10 \times 9 \times 8 \times \cancel{7} \times \cancel{6} \times \cancel{5} \times \cancel{4} \times \cancel{3} \times \cancel{2} \times \cancel{1}}{3 \times 2 \times 1 \times \cancel{7} \times \cancel{6} \times \cancel{5} \times \cancel{4} \times \cancel{3} \times \cancel{2} \times \cancel{1}}$$

$$= \frac{10 \times 9 \times 8}{3 \times 2 \times 1} = \frac{720}{6} = 120 \text{ different ways.}$$

A faster way to think of this is by using the counting principle. Start with the fact that there are 3 different people to be chosen from a total of 10. There are 10, then 9, then 8 choices for the 3 people. Divide this result by the number of permutations, or orders, of the 3 people since order does not matter in combinations. In this way, the number of combinations can be found just using $\frac{10 \times 9 \times 8}{3 \times 2 \times 1} = \frac{720}{6} = 120$.

Practice 6

Practice makes perfect when enhancing your skills with counting problems. Work with the next set of practice problems to help in this area. Answers and explanations are at the end of the chapter.

21. If there are 5 shirts, 4 pairs of pants, and 3 pairs of shoes, how many different outfits can be made using 1 shirt, 1 pair of pants, and 1 pair of shoes?

22. Find $_4P_4$.

23. How many different arrangements of 5 students can be made in a row of 3 desks?

24. Find $_5C_3$.

25. How many different 4-person committees can be formed from a total of 8 people?

SUMMARY

To review, here are five main ideas about statistics that you should take from this chapter:

- The arithmetic mean is the average of a set of numbers.
- The median is the middle number in an ordered list.
- The mode is the number that appears most often.
- The range is the difference between the highest and lowest values in the set.
- Being able to read and interpret data in graphs is an important skill to have; some of the most common graphs used are bar graphs, line graphs, circle graphs, scatter plots, and histograms.

Here are five key concepts about probability that you should take from this chapter:

- The probability of an event (E) is P(E) =
  $$\frac{\text{the number of ways event E can occur}}{\text{the total number of possible outcomes}}$$
- In probability, the word "or" is a key word for addition, and the word "and" is a key word for multiplication.
- There are many different types of counting problems. In order to work with them, multiply out the total number of choices you have to find the total number of possible outcomes.
- A permutation is a special type of counting problem where the order is important and uses the formula $_nP_r = \dfrac{n!}{(n-r)!}$.
- A combination is also a special type of counting problem where the order does not matter and uses the formula
  $$_nC_r = \frac{n!}{r!(n-r)!}.$$

Practice Answers and Explanations

1. C

$$\frac{9 + 7 + 8 + 14 + 22 + 12}{6} = 12$$

2. E

The number 33 appears the most.

3. A

$123 - 96 = 27$

4. B

$5 + 8 = 13$

5. D

6. C

A circle graph shows parts of a whole. The percents must add up to 100%.

7. B

A line graph shows a trend over a period of time.

8. A

A bar graph is used to compare different things.

9. $7.00

The point directly above 5 years on the x-axis is also directly to the right of $7.00 on the y-axis. A person with 5 years experience makes $7.00 per hour.

10. 59

The range is the largest – smallest values in the table. $69 - 10 = 59$.

11. 91 – 100

Because the tallest bar is the interval 91 – 100, then the most grades occur in that interval.

12. $\dfrac{1}{6}$

There is 1 side with a 6, out of 6 possible sides to a die.

13. $\frac{1}{2}$

There is 1 section labeled with a B; however, this section is twice as big as the other two sections. Since it takes up $\frac{1}{2}$ of the spinner, the probability is $\frac{1}{2}$.

14. $\frac{16}{52}$ or $\frac{4}{13}$

The probability of a 2 is $\frac{4}{52}$ and the probability of a face card is $\frac{12}{52}$. Remember, P (A or B) = P(A) + P (B) - P (A and B). Since these events are mutually exclusive, the probability is $\frac{4}{52} + \frac{12}{52} = \frac{16}{52} = \frac{4}{13}$.

15. $\frac{3}{6}$ or $\frac{1}{2}$

The probability of rolling a 3 is $\frac{1}{6}$ and the probability of rolling an odd number is $\frac{3}{6}$. Remember, P (A or B) = P(A) + P (B) - P (A and B). Since 3 is an odd number, the events are mutually inclusive. Therefore, the probability is $\frac{1}{6} + \frac{3}{6} - \frac{1}{6} = \frac{3}{6} = \frac{1}{2}$.

16. $\frac{0}{13}$ or 0

Since there are no pink balls in the bag, the event is impossible so the probability is 0.

17. True

With replacement, the P(king, queen) = $\frac{4}{52} \times \frac{4}{52} = \frac{16}{2704} = \frac{1}{169}$.

18. False

P(5, tails) = $\frac{1}{6} \times \frac{1}{2} = \frac{1}{12}$.

19. False

Without replacement, the P(king, queen) = $\frac{4}{52} \times \frac{4}{51} = \frac{16}{2652}$.

20. True

With replacement, the P(white, blue) $= \frac{4}{12} \times \frac{5}{11} = \frac{20}{132} = \frac{5}{33}$.

21. 60

Use the counting principle to find the total number of possibilities. Multiply the choices together: $5 \times 4 \times 3 = 60$ different outfits.

22. 24

$_4P_4 = 4 \times 3 \times 2 \times 1 = 24$.

23. 60

Since the order is important, this problem is a permutation. Arranging 5 students in a row with 3 desks can be expressed as $_5P_3 = 5 \times 4 \times 3 = 60$.

24. 10

$_5C_3 = \frac{5 \times 4 \times 3}{3 \times 2 \times 1} = \frac{60}{6} = 10$.

25. 70

Choosing a 4 person committee from 8 people can also be expressed as $_8C_4 = \frac{8 \times 7 \times 6 \times 5}{4 \times 3 \times 2 \times 1}$.

CHAPTER 6 TEST

Try the following questions to test your knowledge of statistics and probability. Answer explanations are provided at the conclusion of the chapter to help you assess your understanding of the concepts.

1. The daily snowfall amounts, in inches, for a certain city over a period of 5 days was 12, 2, 1, 0, and 5. What is the mean snowfall amount in this period, in inches?

    (A) 0　(B) 1　(C) 2　(D) 4　(E) 6

2. Five different students working at summer jobs make the following amounts per hour: $6.50, $6.00, $7.25, $5.50, and $7.00. What is the range in pay per hour?

    (A) $1.00　(B) $1.50　(C) $1.75　(D) $2.25　(E) $3.25

3. Given the following set of data, which of the following is true?

   {3, 5, 11, 12, 17, 18, 5, 20, 5, 22, 3}

   I. The mean is equal to the mode.

   II. The median is greater than the mean.

   III. The mode is less than the range.

   (A) I only   (B) II only   (C) III only

   (D) II and III only   (E) I, II, and III

Use the graph below to answer questions 4 and 5.

The graph below shows the number of automobiles sold at a dealership by salesperson.

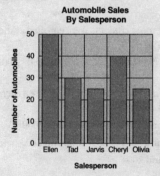

4. According to the bar graph above, how many automobiles were sold by Tad?

   (A) 25   (B) 30   (C) 35   (D) 40   (E) 50

5. What is the ratio of the number of automobiles sold by Jarvis to the number of automobiles sold by Ellen?

   (A) 1:2   (B) 2:3   (C) 3:4   (D) 1:1   (E) 2:1

Use the graph below to answer questions 6 and 7.

The graph below shows the enrollment at a college from 1998 to 2004.

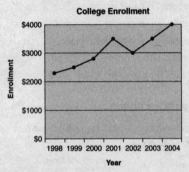

6. How many more students were enrolled at the college in 2004 than in 1999?

   (A) 500   (B) 1,000   (C) 1,500   (D) 2,000   (E) 2,500

7. Between which two years did the greatest increase in enrollment occur?

   (A) 1998 and 1999   (B) 1999 and 2000

   (C) 2000 and 2001   (D) 2002 and 2003

   (E) 2003 and 2004

Use the graph below to answer questions 8 and 9.

The following circle graph shows the breakdown of summer activities of certain students.

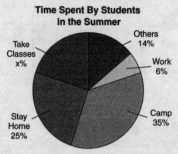

Time Spent By Students in the Summer

8. According to the graph, what percent of students attended classes during the summer?

   (A) 1% (B) 10% (C) 15% (D) 20% (E) 25%

9. If 300 students participated in this study, how many either worked or went to a camp?

   (A) 41 (B) 82 (C) 123 (D) 143 (E) 105

10. What type of correlation is shown in the scatter plot below?

   (A) positive (B) negative (C) no correlation
   (D) choices A, B, and C (E) none of these

11. The following scatter plot shows the relationship between the hours spent studying and the grade on the exam.

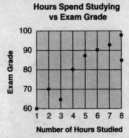

Hours Spend Studying vs Exam Grade

According to the graph, how much time did a person spend studying if she received a 90 on the exam?

(A) 4   (B) 5   (C) 6   (D) 7   (E) 8

1 2. The stem and leaf plot below shows the years of employment of 12 workers at a company.

stem	leaf
1	2, 3, 4, 5
2	4, 4, 5, 8, 9
3	1, 1, 2

According to the chart, what is the greatest number of years of employment for these workers?

(A) 20   (B) 29   (C) 31   (D) 32   (E) 44

13. The histogram below shows the results after students in a class were measured for their heights.

Heights of Students in a Class

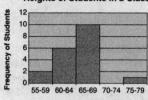

According to the graph, how many total students were measured?

(A) 10 (B) 17 (C) 18 (D) 19 (E) 21

14. Evaluate $_7P_3$.
   (A) 10 (B) 21 (C) 35 (D) 120 (E) 210

15. Evaluate $_{10}C_2$.
   (A) 10 (B) 20 (C) 45 (D) 90 (E) 100

16. At an ice cream shop, Kerensa can choose from 8 different flavors of ice cream, 3 different toppings, and 2 types of cones. If she selects 1 flavor of ice cream, one topping, and one type of cone, how many different combinations can she make?

   (A) 12 (B) 13 (C) 24 (D) 36 (E) 48

17. A bag contains 6 red, 4 blue, 10 orange, and 3 yellow candies. What is P(red or yellow)?

   (A) $\frac{3}{23}$ (B) $\frac{23}{23}$ (C) $\frac{10}{23}$ (D) $\frac{9}{23}$ (E) $\frac{9}{46}$

18. A coin is tossed, the result is noted, and then the coin is tossed again. What is P(heads, heads)?

    (A) $\frac{1}{4}$ (B) $\frac{1}{2}$ (C) $\frac{2}{3}$ (D) 1 (E) none of these

19. A cooler of cold drinks contains 6 bottles of cola, 8 bottles of orange drink, 4 bottles of root beer, and 3 bottles of water. If a drink is selected, not replaced, and then another selected, what is P(orange drink, cola)?

    (A) $\frac{48}{441}$ (B) $\frac{4}{35}$ (C) $\frac{14}{41}$ (D) $\frac{10}{41}$ (E) $\frac{40}{441}$

20. How many ways can 2 students be selected from 7 to be representatives on student council?

    (A) 9 (B) 14 (C) 21 (D) 42 (E) 84

Answers and Explanations

1. D

The mean can be found by finding the sum of the numbers in the set, and then by dividing by the total number of elements in the set. The sum of the set is 20. Since there are 5 numbers, divide 20 by 5 to get a mean of 4. Choice (A) is the lowest number in the list, not the mean. Choice (B) is too small to be the mean of the set. Choice (C) is the median in the set, but not the mean. Choice (E) is too large to be the mean of the set.

2. C

The range in pay can be found by subtracting the smallest amount from the largest amount. $7.25 - 5.50 = $1.75.

3. C

For this set of data, the mean is 11, the median is 11, the mode is 5, and the range is 19. Therefore, the only true statement is statement III—the mode is less than the range: 5 < 19. Choice (A) is incorrect because the mean is 11 and the mode is 5, which are not equal.

Choice (B) is incorrect because both the median and mean are 11; the median is not greater than the mean, they are equal. Thus, choices (D) and (E) are incorrect because statement I and statement II are false.

4.   B

Since the height of Tad's bar reaches 30 on the vertical axis, therefore he sold 30 automobiles.

5.   A

Jarvis sold 25 automobiles and Ellen sold 50. Therefore, the ratio is 25 : 50, which can be simplified to 1 : 2 when each number is divided by 25.

6.   C

The enrollment at the college in 1999 was 2,500; the enrollment in 2004 was 4,000. To find the difference, subtract: 4,000 − 2,500 = 1,500.

7.   C

The greatest increase occurs between the years 2000 and 2001. Here, the enrollment went from 2,800 to 3,500, an increase of about 700 students.

8.   D

Since the percents in a circle graph need to add to 100%, add the known percents and subtract that sum from 100: 14% + 6% + 35% + 25% = 80%; 100% - 80% = 20%.

9.   C

Because 6% worked and 35% went to a camp, add 6% and 35% to get 41%. Now find 41% of 300 by using the proportion $\frac{x}{300} = \frac{41}{100}$. Cross-multiply to get $100x = 12,300$, and divide each side of the equation by 100 to get $x = 123$. Recall from chapter 5 that you can also multiply $0.41 \times 300 = 123$ to find the percent of a number.

10.   B

Since the series of points in the graph have a trend that decreases as you move to the right on the horizontal axis, this is an example of a negative correlation.

11. C

First, find 90 on the vertical axis. Then, look directly to the right until you get to the point at that height. This point is directly above the number 6 on the horizontal axis. Therefore, a person who received a 90 studied for 6 hours.

12. D

The greatest number of years can be found by looking for the greatest stem with the greatest leaf. The greatest stem is 3; the largest leaf in this stem is 2. Therefore, the greatest amount of years is 32.

13. D

To find the total number of students, add the heights of each of the bars. $2 + 6 + 10 + 0 + 1 = 19$.

14. E

The expression $_7P_3 = 7 \times 6 \times 5 = 210$.

15. C

The expression $_{10}C_2 = \dfrac{10 \times 9}{2 \times 1} = \dfrac{90}{2} = 45$.

16. E

Use the counting principle to multiply the total number of choices together: $8 \times 3 \times 2 = 48$ different combinations.

17. D

There are a total of 23 candies in the bag. Because there are 6 red and 3 yellow, the probability of P(red or yellow) is $\dfrac{6}{23} + \dfrac{3}{23} = \dfrac{9}{23}$.

18. A

The probability of getting heads on the coin is always $\dfrac{1}{2}$. Therefore, the probability of getting heads first, and then heads again is $\dfrac{1}{2} \times \dfrac{1}{2} = \dfrac{1}{4}$.

19. B

There are a total of 21 bottles in the cooler. If an orange drink is removed and not replaced, then there are only 20 bottles left in the cooler the second time a drink is chosen. Therefore, the P(orange drink, cola) is equal to $\frac{8}{21} \times \frac{6}{20} = \frac{48}{420} = \frac{4}{35}$.

20. C

This problem is a combination situation. Since you are looking for 2 students to be selected from 7 and the order does not matter, evaluate $_7C_2 = \frac{7 \times 6}{2 \times 1} = \frac{42}{2} = 21$.

Powers and Roots

BUILDING BLOCK QUIZ

This Building Block Quiz will review some of the statistics and probability topics discussed in chapter 6, and will preview the essential elements of powers and roots presented in this chapter. Use the results of this quiz to help you direct your focus as you work through the chapter.

1. What is the median of the following set of data?

   {32, 44, 68, 24, 39, 32, 50}

   (A) 24   (B) 32   (C) 39   (D) 44   (E) 68

2. When tossing a coin two times, what is the probability of getting heads both times?

   (A) $\frac{1}{4}$ (B) $\frac{1}{3}$ (C) $\frac{1}{2}$ (D) $\frac{1}{8}$ (E) $\frac{1}{16}$

3. How many different 3-digit numbers can be formed using the digits 0 through 9?

   (A) 10   (B) 81   (C) 100   (D) 729   (E) 1,000

4. $4^3 =$

   (A) 7   (B) 12   (C) 16   (D) 48   (E) 64

5. $2^4 \times 4^2 =$

   (A) 2^2 (B) 2^6 (C) 4^6 (D) 2^8 (E) 2^{16}

6. $\sqrt{289} =$

   (A) 2.89 (B) 8.5 (C) 17 (D) 144.5 (E) 289

7. $5^{-2} =$

   (A) −10 (B) −25 (C) $\dfrac{1}{10}$ (D) $\dfrac{-1}{10}$ (E) $\dfrac{1}{25}$

8. $9^{\frac{3}{2}}$

   (A) 3 (B) 9 (C) 13.5 (D) 27 (E) 81

9. Simplify $\sqrt{\dfrac{9}{16}}$.

   (A) $\dfrac{9}{16}$ (B) $\dfrac{3}{4}$ (C) $\dfrac{4}{3}$ (D) 3 (E) 4

10. $6.35 \times 10^{-3} =$

   (A) 0.635 (B) 0.0635 (C) 0.00635

   (D) 0.000635 (E) 6.35

Answers and Explanations

1. C

To find the median of a set of data, put the numbers in order from least to greatest, and find the middle number in the list. The numbers in order are {24, 32, 32, 39, 44, 50, 68}. The middle number in this list is 39.

2. A

This question tests the concept of independent events of probability from chapter 6. To find the probability of both events happening, multiply the probability of getting heads each time: $\dfrac{1}{2} \times \dfrac{1}{2} = \dfrac{1}{4}$.

3. E

This problem tests your knowledge of the number of arrangements, or permutations, that can be made. Since there are three digits and 10 choices for each digit, multiply $10 \times 10 \times 10 = 1,000$.

4. E

Questions 4 and 5 test your knowledge of exponents. Use the base number of 4 as a factor three times: $4 \times 4 \times 4 = 64$.

5. D

Using order of operations, evaluate the exponents first and then multiply the results: $2^4 = 2 \times 2 \times 2 \times 2 = 16$ and $4^2 = 4 \times 4 = 16$. $16 \times 16 = 256$. Since this is not an answer choice, evaluate the choices to find the one equal to 256. $2^8 = 256$.

6. C

This question addresses the concept of perfect squares. Since $17 \times 17 = 289$, then $\sqrt{289} = 17$.

7. E

In this question, you are asked to evaluate a negative exponent. To evaluate a negative exponent, take the reciprocal of the base, make the exponent positive, and then evaluate the positive exponent.

$5^{-2} = \dfrac{1}{5^2} = \dfrac{1}{5 \times 5} = \dfrac{1}{25}$.

8. D

This question tests the concept of fractional exponents. In a fractional exponent, the numerator is the power and the denominator is the root.

$9^{\frac{3}{2}} = \left(\sqrt{9}\right)^3 = 3^3 = 27$.

9. B

In this problem, you are asked to simplify the square root of a fraction. To do this, first separate the numerator from the denominator and then evaluate the square roots. $\sqrt{\dfrac{9}{16}} = \dfrac{\sqrt{9}}{\sqrt{16}} = \dfrac{3}{4}$. Note that both 9 and 16 were perfect squares and had integers as square roots.

10. C

This question tests the concepts of scientific and standard notation. To convert into standard notation, use the first factor of 6.35 and move the decimal 3 places to the left. Put zeros as place holders where necessary, and note that the decimal moved to the left because the exponent was negative. The number in standard notation is 0.00635.

EXPONENTS

Exponents are a way to write very small and very large numbers in a shortened fashion, which can save both time and space. This chapter will review the applications of exponents, and will provide various practice exercises to help you refine your skills. Some of the topics covered include exponents as powers, the laws of exponents, negative exponents, fractional exponents (otherwise known as radicals and roots), and scientific notation.

Exponents as Powers

Using exponents is a way to write repeated multiplication more efficiently. They are used when a factor being multiplied several times is the same.

MATH SPEAK

The **exponent** or **power** is the total number of times a base is used as a factor.

The number or expression that appears below the exponent is called the *base number* or *base expression*. The exponent is always written in smaller type and sits a little above and to the right of the base number. In the numerical expression 6^4, 6 would be considered the *base number* and 4 the *exponent*. As stated above, the exponent in the expression tells you how many times to use the base number as a factor.

For example, in the expression 6^4, 6 should be used as a factor 4 times.

$$6^4 = 6 \times 6 \times 6 \times 6 = 1{,}296$$

When an expression is written out showing all of its factors, as above, this is called the *expanded form* of the expression.

Sometimes a number or expression will have no exponent. If there is not an exponent written with a base number or variable, the exponent is equal to 1. For example, the numerical expression 6 actually means 6^1.

In another special case, any base expression to the 0 power is equal to 1. For example, the expressions 5^0, 101^0, and $12{,}045^0$ all equal 1.

When simplifying expressions containing exponents, evaluating the exponent(s) is an early step in the order of operations. In fact, you would evaluate exponents directly after simplifying within any parentheses or grouping symbols in your expression.

FLASHBACK

Remember the acronym **PEMDAS** from chapter 1 that helps you remember the correct order of operations: <u>P</u>arentheses, <u>E</u>xponents, <u>M</u>ultiplication and <u>D</u>ivision, <u>A</u>ddition and <u>S</u>ubtraction.

Examples:

Evaluate $(-3) - (2)^3 + 1$

When simplifying the expression $(-3) - (2)^3 + 1$, there are no expressions contained within the parentheses that need to be simplified, so your first step is to evaluate the exponent of 3 on the base number of 2: $2^3 = 2 \times 2 \times 2 = 8$. Notice that the exponent only belongs to the base it is directly next to. The expression becomes $-3 - 8 + 1$. Perform subtraction next to get $-11 + 1$, and then add to get a final result of -10.

Evaluate $3^4 \times 4^3$

First, evaluate the exponents using expanded form to help visualize the operation. $3^4 = 3 \times 3 \times 3 \times 3 = 81$ and $4^3 = 4 \times 4 \times 4 = 64$. Now, multiply: $81 \times 64 = 5,184$.

Evaluate $(7)^2, (-7)^2,$ and $-(7)^2$

The first expression $(7)^2$ is equal to $7 \times 7 = 49$.

The second expression $(-7)^2$ is equal to $(-7) \times (-7) = 49$.

However, the third expression $-(7)^2$ is equal to $-[(7) \times (7)] = -49$. Since the negative is not within the parentheses, it is not raised to the second power. The expression reads as the opposite of 7 to the second power. Using the correct order of operations leads to an answer of -49, instead of 49.

Practice 1

Use the bank of numbers below to answer each question. Answers and explanations are located at the end of the chapter.

10   108   4   14   16   -15   25   32   17   21   81

1. The expression 2^5 is equal to _____.

2. In the expression 16^{14}, the base number is _____.

3. In the expression 7^4, seven should be used as a factor _____ times.

4. The expression $(-4)^2 + 5^0$ is equal to _____.

5. The expression $2^2 \times 3^3$ is equal to _____.

Multiplying Like Bases

When working with exponents there are shortcuts to simplifying expressions. Understanding expanded form will help you comprehend and apply these different laws of exponents.

One of the patterns that occurs when you are multiplying exponents and the base numbers are the same. These are called *like bases*. In the example $2^3 \times 2^5$, 2 is the common base number, or the like base. When writing the expressions out in expanded form, 2^3 is equal to $2 \times 2 \times 2$ and 2^5 is equal to $2 \times 2 \times 2 \times 2 \times 2$. So the expression $2^3 \times 2^5$ then becomes $(2 \times 2 \times 2) \times (2 \times 2 \times 2 \times 2 \times 2)$. Since 2 is used as a factor 8 times, this expression can be rewritten as 2^8. Notice that the exponents of 3 and 5 were added together to get an exponent of 8. Thus, when multiplying like bases, add the exponents.

Dividing Like Bases

A pattern also occurs when dividing exponents where the base expressions are the same. For example, in the expression $\dfrac{5^6}{5^4}$, 5 is the common base for each expression. Recall that the fraction bar means division. Write both the numerator and denominator in expanded form and cancel out the same number of factors from each.

The expression then becomes $\dfrac{5^6}{5^4} = \dfrac{5 \times 5 \times 5 \times 5 \times 5 \times 5}{5 \times 5 \times 5 \times 5} =$

$\dfrac{\cancel{5} \times \cancel{5} \times \cancel{5} \times \cancel{5} \times 5 \times 5}{\cancel{5} \times \cancel{5} \times \cancel{5} \times \cancel{5}}$. Notice that 4 pairs of factors of 5 could be

canceled from the numerator and denominator, leaving only 5×5 in the numerator. The simplified expression is then equal to 5^2, or 25. As in the generalization for multiplication, compare the original exponents with the result to determine the rule. Since the original exponents were 6 and 4, and the result was an exponent of 2, the exponents were subtracted when the like bases were divided.

> ### REMEMBER THIS
> To **multiply** with like bases, add the exponents: $a^m \times a^n = a^{m+n}$.
> To **divide** with like bases, subtract the exponents: $= \dfrac{a^m}{a^n} = a^{m-n}$.

Practice 2

Answer true or false for the following questions. Answers and explanations are located at the end of the chapter.

6. **T F** The expression $\frac{6^7}{6^2}$ is equivalent to 6^9.

7. **T F** When multiplying $10^2 \times 10^4$, the result is equivalent to 10^6.

8. **T F** The expression $2^5 \times 5^2$ is equivalent to 10^7.

9. **T F** The expression $\frac{5^5 \times 5^3}{5^4}$ simplifies to 5^4.

Operations with Powers

Raising a Power to a Power

Another situation you might encounter is raising an exponent, or power, to another power. An example of this is the expression $(2^3)^4$. In this type of problem, use expanded form to help visualize your operation. Since the exponent of 4 is outside of the parentheses, it implies that $(2^3)^4 = (2^3) \times (2^3) \times (2^3) \times (2^3)$. In other words, the base of 2^3 is being used as a factor 4 times. By adding the exponents of the like bases as explained above, the result is 2^{12}. Since raising the exponents of 3 to the 4th power gave a resulting exponent of 12, you can see that the original exponents were multiplied together in order to simplify the expression.

Raising a Product or Quotient to a Power

When a base expression contains more than just one variable or constant and is then raised to another power, this is called raising a product to a power. An example is $(3x^4)^3$, where the entire expression $3x^4$ is raised to the third power. The good news is that the same principle mentioned above also applies to this example. The catch is that you need to remember to multiply each base number or variable's exponent by 3. Keep in mind that the bases here are 3 and x. $(3x^4)^3$ becomes $3^3 x^{4 \times 3}$, which simplifies to $127x^{12}$ using the power to a power multiplication rule.

The product to a power principle is also used if an expression contains a fraction. In this case both numerator and denominator are raised to the exponent using these same rules. For example, in the expression $\left(\frac{3^2}{2} \right)^3$, the exponent of 3 needs to be applied to the base in the numerator of 3^2 and the base in the denominator of 2. The simplified expression would be $\left(\frac{3^2}{2} \right)^3 = \frac{(3^2)^3}{(2)^3} = \frac{3^6}{2^3} = \frac{729}{8}$. The rules for raising a power to a power and raising a product to a power were both used to simplify.

REMEMBER THIS!

To raise an exponent to another power, multiply the exponents: $(a^m)^n = a^{m \times n}$.

To raise a product to a power, raise each base number and/or variable to that power: $(a^m b^n)^t = a^{m \cdot t} b^{n \cdot t}$. To raise a fraction to a power, raise both the numerator and denominator to that exponent and simplify the expression: $\left(\frac{a}{b} \right)^m = \frac{a^m}{b^m}$.

Practice 3

Match each question in the left column with the correct answer choice from the right column. Answers and explanations are located at the end of the chapter.

10. $(8^3)^5 =$ A. $512x^{12}$

11. $(8x^5)^3 =$ B. 8^{15}

12. $(8^4)(8^2)^3 =$ C. 8^6

13. $\left(\frac{8^3}{8^2} \right)^6 =$ D. $512x^{15}$

14. $(8x^4)^3 =$ E. 8^{10}

Negative Exponents

Working with negative exponents can be a challenge. A common misconception is that the negative sign on the exponent makes the base number negative—not the case! A base expression raised to a negative exponent is equal to the *reciprocal* of the base raised to a positive exponent. To evaluate a negative exponent, first take the reciprocal of the base. Then, make the exponent positive and evaluate as is.

For example, $5^{-3} = \frac{1}{5^3} = \frac{1}{125}$. An exponent that was negative in the numerator will be positive in the denominator. In the same fashion, an exponent that was negative in the denominator will be positive in the numerator. Take a look at the following examples:

Simplify $\frac{2^{-3}}{4^{-2}}$.

First, rewrite the expression with positive exponents by taking the reciprocal of any base with a negative exponent. In this case, the expression would become $\frac{4^2}{2^3}$. Notice that when the bases switched between the numerator and denominator, the sign of the exponent also switched. Now, simplify by evaluating the exponents:

$\frac{4^2}{2^3} = \frac{4 \times 4}{2 \times 2 \times 2} = \frac{16}{8} = 2$.

Simplify $\frac{5^{-2} \times 2^2}{2^{-3}}$.

First, rewrite the expression with positive exponents by taking the reciprocal of any base with a negative exponent. The expression then becomes $\frac{2^2 \times 2^3}{5^2}$. Multiply in the numerator by adding the exponents: $\frac{2^5}{5^2}$. Simplify the expression by evaluating the exponents:

$\frac{2^5}{5^2} = \frac{32}{25}$.

> ## REMEMBER THIS!
>
> To simplify an expression with negative exponents, remember that $a^{-m} = \dfrac{1}{a^m}$.

Try another example of a problem dealing with negative exponents and a variable.

If $x^{-2} = \dfrac{1}{9}$, what is the value of x?

Since 3^2 is equal to 9, then 3^{-2} is equal to $\dfrac{1}{3^2} = \dfrac{1}{9}$. Therefore, $x = 3$.

Practice 4

Try the following set of problems to assess your understanding of negative exponents. Answers and explanations are located at the end of the chapter.

15. Simplify $\dfrac{3^{-4}}{4^{-1}}$.

16. Simplify $\dfrac{6^{-2} \times 4^2}{6^2}$.

17. If $x^{-5} = \dfrac{1}{32}$, what is the value of x^2?

Scientific Notation

A common application of exponents is **scientific notation**. Scientific notation is often used to write very large or very small numbers more efficiently. To write a number in scientific notation, place a decimal within the non-zero digits of the number to create a number between 1 and 10. Then, multiply this number by a factor of 10, where the exponent to the 10 is the number of decimal places the decimal point has moved.

For example, to change the number 6,400,000 to scientific notation, write the non-zero digits as a number between 1 and 10 and drop the zeros. The number becomes 6.4. Then multiply this number by 10 raised to an exponent; the exponent is the number of places the decimal had to move to get between the 6 and the 4. By counting from the ones place of 6,400,000 to the space between 6 and 4, the decimal moved 6 places to the left.

$$6{,}400{,}000$$
6 places

Therefore, 6,400,000 written in scientific notation is equal to 6.4×10^6.

Because the number in the example above was greater than 1, the exponent on 10 in scientific notation was positive. If the number being converted is less than 1, the exponent on 10 will be negative. For instance, take the number 0.00000045 and follow the same procedure as above. Write the non-zero digits as a number between 1 and 10 and drop the zeros to get 4.5. Now, multiply this value by 10 raised to a power, where the exponent is the total number of places that the decimal point moved. Since the number is less than 11, it moved 7 places to the right and the exponent is −7.

$$0.00000045$$
7 places

Therefore, $0.00000045 = 4.5 \times 10^{-7}$.

You can also convert from scientific notation to standard notation. If converting 2.3×10^4 to standard notation, write the first factor of 2.3 and then move the decimal 4 places to the right. Add zeros as place holders where necessary. Thus, the number becomes 23,000 in standard notation. Note that if the exponent were negative in scientific notation, the decimal would move that number of places to the left instead of the right. Remember that a negative exponent indicates a number less than 1. $5.4 \times 10^{-3} = 0.0054$ in standard notation.

REMEMBER THIS!

To convert a number to **scientific notation**, first use the non-zero digits to make a number between 1 and 10. Then, multiply by a factor of 10 to a power, where the power is the number of places the decimal moved. If the number is less than 1, the exponent is negative; if the number is greater than one, the exponent is positive.

Practice 5

Answer true or false for each of the following. Answers and explanations are located at the end of the chapter.

18. The number 9,000,000 is written as 9.0×10^9 in scientific notation.

19. The number 0.00086 is written as 8.6×10^{-4} in scientific notation.

20. The number 6.5×10^5 is equal to 650,000 in standard notation.

ROOTS AND FRACTIONAL EXPONENTS

Perfect Squares

The study of roots and fractional exponents begins with the concept of perfect squares. Perfect squares are numbers that are the result of multiplying two of the same integers together. Some examples are:

$1 \times 1 = 1$ $2 \times 2 = 4$ $3 \times 3 = 9$ $4 \times 4 = 16$ $5 \times 5 = 25$

Therefore, the perfect squares listed above are 1, 4, 9, 16, and 25. Other perfect squares can be generated in the same way. In each of these cases, the pattern was the same; $1^2 = 1$, $2^2 = 4$, $3^2 = 9$, $4^2 = 16$, $5^2 = 25$. The square of the integer was equal to the perfect square.

> **MATH SPEAK**
>
> The symbol for a **square root** is $\sqrt{}$, and this symbol is known as the radical sign. The number under the **radical sign** is called the **radicand**.

Finding the square root of a number, or a radicand, is the opposite of finding the square of a number. To find the square root of a number, find the integer that you would multiply by itself to equal the number. For example, since $4^2 = 4 \times 4 = 16$, then $\sqrt{16} = 4$. Four is the square root of 16. In addition, since $10^2 = 10 \times 10 = 100$, then $\sqrt{100} = 10$. Ten is the square root of 100.

The root of a number can also be written as an exponent. However, when the root of a base number is involved, the exponent will be in fraction form. For any square root, the fractional exponent will always be $\frac{1}{2}$, where the root (or index) is 2 and the power is 1. For example:

$$25^{\frac{1}{2}} = \sqrt{25} = 5, \ 81^{\frac{1}{2}} = \sqrt{81} = 9, \text{ and } 169^{\frac{1}{2}} = \sqrt{169} = 13.$$

There are additional cases of fractional exponents, other than $\frac{1}{2}$. The numerator of the fraction is the power of the base number, and the denominator is the root of the base number. Because of this fact, fractional exponents can be written as $a^{\frac{x}{y}} = \left(\sqrt[y]{a}\right)^x$. In the case of $4^{\frac{5}{2}}$, take the square root of 4 and then raise that result to the power of 5

$$4^{\frac{5}{2}} = (\sqrt{4})^5 = 2^5 = 32.$$

Square Roots That Are Not Perfect Squares

As you probably could predict, not all square roots are perfect squares. If the number in the radicand is not the product of a number and itself, then the number is irrational.

> **FLASHBACK**
> Revisit chapter 1 to review the different sets of numbers, including **irrational numbers**.

But just like fractions, most radicals can be expressed in a reduced, or simplified form, even if they are not perfect squares. To simplify radicals, look for the perfect square factors of the number in the radical. In other words, find the largest perfect square that divides evenly into your radicand without a remainder. Take, for example, the square root of 8, or $\sqrt{8}$. Since 8 is the product of 4 and 2, then $\sqrt{8} = \sqrt{4} \times \sqrt{2}$. Because the square root of 4 is equal to 2, this expression can be simplified to $2 \times \sqrt{2}$ or $2\sqrt{2}$. Four is the largest perfect square factor of 8, and simplifying this factor reduces the radical to simplest radical form.

For another example, simplify $\sqrt{54}$. Since the largest perfect square factor of 54 is 9, express $\sqrt{54}$ as $\sqrt{9} \times \sqrt{6}$. The square root of 9 is 3, so $\sqrt{54} = 3\sqrt{6}$.

An additional twist to simplifying radicals happens when there is a coefficient, or a number in front of the radical. In this situation, simplify the radicand as explained above, and then multiply the coefficient by any number taken out of the radical. For example, take the expression $5\sqrt{44}$. Since the largest perfect square factor of 44 is 4, the expression becomes $5 \times \sqrt{4} \times \sqrt{11}$. The square root of 4 is 2 so the expression is $5 \times 2 \times \sqrt{11}$, which simplifies to $10\sqrt{11}$.

Always keep in mind that a radical is in simplest form if there are no perfect square factors contained in the radicand.

Practice 6

Express each in simplest form. Answers and explanations are located at the end of the chapter.

21. $\sqrt{144} =$

22. $289^{\frac{1}{2}} =$

23. $16^{\frac{3}{2}} =$

24. $\sqrt{72} =$

25. $2\sqrt{24} =$

Operations With Radicals

Addition and Subtraction

Adding and subtracting radicals is very much like adding and subtracting with fractions—you need to have like terms. In the case of radicals, you must have the same radicand in order to add and subtract, instead of a common denominator. If the terms being combined do not have the same radicand to begin with, simplify each term to see if a common radicand is possible and then only add the terms with this common radicand.

Examples:
$$4\sqrt{3} + 5\sqrt{3} = 9\sqrt{3}$$

Since each term has a common radicand of 3, add the coefficients of 4 and 5 and keep the radical the same.
$$4\sqrt{2} - \sqrt{8} = 4\sqrt{2} - 2\sqrt{2} = 2\sqrt{2}$$

Since the terms did not have the same radicand, the square root of 8 was reduced to $2\sqrt{2}$. Subtract the coefficients and keep the radical to get an answer of $2\sqrt{2}$.
$$\sqrt{6} + 2\sqrt{11}$$

These terms cannot be combined. Each term is in simplest form and there is not a common radicand between them.

> ## REMEMBER THIS!
>
> To add or subtract radicals, be sure to have the same radicand for all terms.
>
> If the radicands are not the same and all are in simplest form, the radicals cannot be combined (i.e. $\sqrt{2} + \sqrt{5}$ cannot be combined).

Multiplication and Division

When multiplying and dividing radicals, no common radicand is necessary. Multiply (or divide) coefficients with coefficients, and radicals with radicals. Use the examples below as a reference:

$$2\sqrt{5} \times 3\sqrt{3} = 6\sqrt{15}$$

The coefficients of 2 and 3 were multiplied to get a coefficient of 6. The radicands of 5 and 3 were multiplied to get a radicand of 15.

$$\frac{15\sqrt{10}}{5\sqrt{2}} = 3\sqrt{5}$$

Because 15 divided by 5 is 3, the coefficient of the result is 3. Since 10 divided by 2 is 5, the radicand of the solution is 5. The final answer is $3\sqrt{5}$.

$$2\sqrt{16} \times 7\sqrt{9} = (2 \times 4) \times (7 \times 3) = 8 \times 21 = 168$$

Since 16 and 9 were both perfect squares, each was converted to its square root and then multiplied by its coefficient. These two results were then multiplied together to get a final answer of 168.

An important thing to note at this point is that any radical in simplest form *cannot* be written with a radical in the denominator of a fraction. To take care of this, you need to do a procedure called **rationalizing the denominator**. Take an example such as $\frac{3}{\sqrt{2}}$. To rationalize the denominator, multiply both the denominator and numerator by $\sqrt{2}$. When doing this, you are really just multiplying the fraction by 1. But performing this simple operation will take the radical out of the denominator by creating a perfect square within the radical sign.

$$\frac{3}{\sqrt{2}} \times \frac{\sqrt{2}}{\sqrt{2}} = \frac{3\sqrt{2}}{\sqrt{4}} = \frac{3\sqrt{2}}{2}$$

Since there are no common factors between the coefficients of 3 and 2 and the denominator is rational, the expression is simplified. Similarly, any radicand that is a fraction needs to be rationalized. In the radical $\sqrt{\frac{4}{3}}$, divide the fraction into numerator and denominator by placing each number under its own radical sign and simplify.

$\sqrt{\frac{4}{3}} = \frac{\sqrt{4}}{\sqrt{3}} = \frac{2}{\sqrt{3}}$. Notice that since 4 is a perfect square, the square root is an integer. Now, rationalize the denominator by multiplying by $\frac{\sqrt{3}}{\sqrt{3}}$. $\frac{2}{\sqrt{3}} \times \frac{\sqrt{3}}{\sqrt{3}} = \frac{2\sqrt{3}}{3}$. This fraction is simplified.

REMEMBER THIS!

A radical is in simplified form if:

- There are no perfect square factors of the radicand other than 1.
- There are no fractions under the radical sign.
- There are no radicals in the denominator of a fraction.

Practice 7

Complete each of the following with the best answer in simplest radical form. Answers and explanations are located at the end of the chapter.

26. The sum of $6\sqrt{5} + 2\sqrt{45}$ is _____.

27. The difference of $12\sqrt{6} - 10\sqrt{6}$ is _____.

28. The product of $2\sqrt{7} \times 12\sqrt{2}$ is _____.

29. The quotient $\dfrac{24\sqrt{6}}{8\sqrt{2}}$ is equal to _____.

30. The expression $\dfrac{7}{\sqrt{2}}$ written is simplest form is _____.

31. The expression $\sqrt{\dfrac{5}{6}}$ written in simplest form is _____

SUMMARY

Problems involving exponents and roots are common both on critical tests and in the real world. In conclusion, here are five things about exponents, powers, and roots, that you should take from this chapter:

- Any nonzero base to the exponent of 1 is equal to itself; any nonzero base to the exponent of 0 is equal to 1.
- When multiplying like bases, add the exponents; when dividing like bases, subtract the exponents.
- Remember that radicals are in simplest form if: they contain no perfect square factors, there are no fractions under the radical, and there are no radicals in the denominator of any fraction.
- When adding and subtracting radicals, be sure to have the same radicand before combining. A common radicand is not necessary for multiplying and dividing.
- Use scientific notation to express very large and very small numbers. This notation is always a number between one and ten, multiplied by an appropriate power of 10.

Practice Answers and Explanations

1. **32**

2. **16**

3. **4**

4. **17**
$(-4)^2 + 5^0 = 16 + 1 = 17.$

5.   108

$2^2 \times 3^3 = 4 \times 27 = 108$.

6.   False

Subtract the exponents of 7 and 2 to get 6^5.

7.   True

When multiplying and the bases are the same, add the exponents.

8.   False

Since the bases are not the same, this problem has to be evaluated using order of operations. The result is $32 \times 25 = 800$, which is not equal to 10^7.

9.   True

Since the bases are the same, add the exponents in the numerator to get $\dfrac{5^8}{5^4}$. Subtract the exponents to complete the division: 5^4.

10.   B

11.   D

12.   E

13.   C

14.   A

15. $\dfrac{4}{81}$

Since there are negative exponents, take the reciprocal of the bases and evaluate using positive exponents. $\dfrac{3^{-4}}{4^{-1}} = \dfrac{4^1}{3^4} = \dfrac{4}{81}$.

16. $\dfrac{1}{81}$

Take the reciprocal of the base with the negative exponent and evaluate using the positive exponents. $\dfrac{6^{-2} \times 4^2}{6^2} = \dfrac{4^2}{6^2 \times 6^2} = \dfrac{4^2}{6^4} = \dfrac{16}{1296} = \dfrac{1}{81}$.

17. 4

Since $x^{-5} = \dfrac{1}{32}$, then $x^5 = 32$. Thus, $x = 2$ because $2^{-5} = \dfrac{1}{2^5} = \dfrac{1}{32}$.

This question asks for x^2, so $2^2 = 4$.

18. False

9,000,000 is equal to 9.0×10^6.

19. True

20. True

21. 12

$12 \times 12 = 144$.

22. 17

$\sqrt{289} = 17$ since $17 \times 17 = 289$.

23. 64

With a fractional exponent, the numerator is the power and the

denominator is the root. Thus, $16^{\frac{3}{2}} = (\sqrt{16})^3 = 4^3 = 64$.

24.

Since $72 = 36 \times 2$, then $\sqrt{72} = \sqrt{36} \times \sqrt{2} = 6\sqrt{2}$.

25. $4\sqrt{6}$

Since $\sqrt{24} = \sqrt{4} \times \sqrt{6} = 2\sqrt{6}$, multiply 2 by $2\sqrt{6}$ to get $4\sqrt{6}$.

26. $12\sqrt{5}$

First, convert $2\sqrt{45}$ into simplified form.

$2\sqrt{45} = 2 \times \sqrt{9} \times \sqrt{5} = 2 \times 3 \times \sqrt{5} = 6\sqrt{5}$. Then, combine like

terms: $6\sqrt{5} + 6\sqrt{5} = 12\sqrt{5}$.

27. $2\sqrt{6}$

Because the terms already have the same radicand, subtract the

coefficients and keep the radical the same. $12\sqrt{6} - 10\sqrt{6} = 2\sqrt{6}$.

28. $24\sqrt{14}$

Multiply the coefficients together, and then the radicals to get the

answer. $2\sqrt{7} \times 12\sqrt{2} = 2 \times 12 \times \sqrt{7} \times \sqrt{2} = 24\sqrt{14}$.

29. $3\sqrt{3}$

Divide coefficient by coefficient, radical by radical.

$$\frac{24\sqrt{6}}{8\sqrt{2}} = \frac{24}{8} \times \sqrt{\frac{6}{2}} = 3\sqrt{3}.$$

30. $\dfrac{7\sqrt{2}}{2}$

To express in simplest form, multiply both the numerator and denominator by $\sqrt{2}$ so there is not a radical left in the denominator.

31. $\dfrac{\sqrt{30}}{6}$

First, express numerator and denominator as separate radicals.

$\sqrt{\dfrac{5}{6}} = \dfrac{\sqrt{5}}{\sqrt{6}}$. Then, multiply both the numerator and denominator by

$\sqrt{6}$ so there is not a radical left in the denominator.

$$\frac{\sqrt{5}}{\sqrt{6}} \times \frac{\sqrt{6}}{\sqrt{6}} = \frac{\sqrt{30}}{6}.$$

CHAPTER 7 TEST

Try the following questions to test your knowledge of powers, roots, and exponents. Assess your progress by using the complete answer explanations that follow the test. Turn back to the appropriate section(s) in the lesson portion of the chapter for more help and clarification of the key concepts.

1. $5^4 =$
   (A) 9   (B) 20   (C) 25   (D) 125   (E) 625

2. $13^0 - (-2)^3 =$
   (A) –7   (B) 1   (C) 7   (D) 9   (E) 21

3. The formula for the surface area of a cube is $SA = 6e^2$, where e represents the length of an edge. What is the surface area, in square inches, of a cube with an edge length of 5 inches?
   (A) 25   (B) 31   (C) 60   (D) 125   (E) 150

4. Multiply $3^2 \times 2^3$.
   (A) 6^5 (B) 6^6 (C) 81 (D) 2^6 (E) 72

5. Simplify $(5^2)^3$.
   (A) 5^1 (B) 5^5 (C) 5^6 (D) 5^8 (E) 5^{12}

6. Simplify $(2y^3)^5$.
   (A) $10y8$ (B) $10y15$ (C) $2y8$ (D) $2y15$ (E) $32y15$

7. Simplify $(6^2)(6^2)^4$.
   (A) 6^2 (B) 6^8 (C) 6^{10} (D) 6^{12} (E) 6^{16}

8. Simplify $\left(\dfrac{4^8}{4^7} \right)^5$.

   (A) 4^5 (B) 4^6 (C) 4^{13} (D) 4^{15} (E) 4^{20}

9. Evaluate $\dfrac{3^{-2}}{2^{-3}}$.

   (A) $\dfrac{8}{9}$ (B) $\dfrac{9}{8}$ (C) 1 (D) –1 (E) $\dfrac{2}{3}$

10. Simplify $\dfrac{5^2 \times 2^{-2}}{5^{-2}}$.

    (A) $\dfrac{1}{4}$ (B) $\dfrac{625}{4}$ (C) $\dfrac{4}{625}$ (D) 5 (E) 4

11. If $x^{-3} = \dfrac{1}{27}$, then what is the value of x?

    (A) 27 (B) 9 (C) 3 (D) $\dfrac{1}{3}$ (E) $\dfrac{1}{27}$

12. The distance between two planets in a solar system is 750,000,000 miles. What is this distance written in scientific notation?

    (A) 0.75×10^7 (B) 7.5×10^7 (C) 7.5×10^8
    (D) 0.75×10^6 (E) 75.0×10^8

13. What is the value of 6.02×10^{-6} in scientific notation?

    (A) 6,020,000 (B) 602,000,000 (C) 0.0000602
    (D) 0.00000602 (E) 0.000000602

14. $25^{\frac{3}{2}} =$

    (A) 25 (B) 37.5 (C) 50 (D) 125 (E) 150

15. Express in simplest radical form: $\sqrt{48}$.

    (A) $\sqrt{24}$ (B) $4\sqrt{3}$ (C) $3\sqrt{4}$ (D) $2\sqrt{6}$ (E) $6\sqrt{4}$

16. $\left(9\sqrt{81}\right)^2 =$

    (A) 18 (B) 81 (C) 162 (D) 1458 (E) 6561

17. $\sqrt{28} + 3\sqrt{7} =$

    (A) $\sqrt{38}$ (B) $7\sqrt{7}$ (C) $3\sqrt{35}$ (D) $4\sqrt{7}$ (E) $5\sqrt{7}$

18. $\left(8\sqrt{3}\right)\left(4\sqrt{2}\right) =$

    (A) $12\sqrt{5}$ (B) 32 (C) $32\sqrt{5}$
    (D) $32\sqrt{6}$ (E) none of these

19. Express $\dfrac{11}{\sqrt{7}}$ in simplest radical form.

    (A) 4 (B) $11\sqrt{7}$ (C) $\dfrac{11\sqrt{7}}{7}$

    (D) $\dfrac{7}{\sqrt{7}}$ (E) $\dfrac{11\sqrt{7}}{77}$

20. Express $\sqrt{\dfrac{12}{7}}$ in simplest radical form.

(A) $12\sqrt{7}$ (B) $7\sqrt{12}$ (C) $\dfrac{12\sqrt{7}}{7}$

(D) $\dfrac{84}{\sqrt{7}}$ (E) $\dfrac{2\sqrt{21}}{7}$

Answers and Explanations

1.   E

Use the base of 5 as a factor four times: $5^4 = 5 \times 5 \times 5 \times 5 = 625$.

2.   D

Evaluate the exponents first and then subtract. $13^0 = 1$ and $(-2)^3 = (-2)(-2)(-2) = -8$. The expression becomes $1 - (-8) = 1 + 8 = 9$.

3.   E

Substitute the given value of e into the formula $SA = 6e^2$. $SA = 6 \times 5^2$. Evaluate the exponents first, then multiply by 6: $SA = 6 \times 25 = 150$ square inches.

4.   E

Evaluate the exponents first. $3^2 = 9$ and $2^3 = 8$. Then, multiply: $9 \times 8 = 72$.

5.   C

Since you are raising a power to another power, multiply the exponents to get 5^6.

6.   E

Since you are raising a product to a power, raise each part of the base expression to the power of 5. $2^5 \times y^{3 \times 5} = 32y^{15}$.

7.   C

First, simplify the second set of parentheses by multiplying the exponents to get 6^8. Since you are now multiplying like bases, add the exponents: $6^2 \times 6^8 = 6^{2+8} = 6^{10}$.

8. A

First, raise the entire fraction to the power of 5 by multiplying the

exponents. The expression becomes $\frac{4^{40}}{4^{35}}$. Then, simplify the

expression by subtracting the exponents to get 4^5. In this particular
problem, you could also evaluate within the parentheses first by
subtracting the exponents. The expression would then become $(4^1)5$,
which also equals 4^5.

9. A

Because each of the bases have negative exponents, take the
reciprocal of each base and write each base with a positive exponent.

The expression then becomes $\frac{2^3}{3^2} = \frac{2 \times 2 \times 2}{3 \times 3} = \frac{8}{9}$.

10. B

Write the expression with positive exponents by taking the reciprocal of
any bases with negative exponents. The expression then becomes

$\frac{5^2 \times 5^2}{2^2}$. Now, simplify by multiplying out the numerator and

denominator. $\frac{5^{2+2}}{2^2} = \frac{5^4}{2^2} = \frac{5 \times 5 \times 5 \times 5}{2 \times 2} = \frac{625}{4}$.

11. C

Because $3^3 = 27$, then $3^{-3} = \frac{1}{27}$. Therefore, $x = 3$.

12. C

First, write the non-zero digits as a number between 1 and 10; 7.5.
Then, multiply by a factor of ten to a power that represents the number
of places the decimal moved to get between the 7 and 5, which is 8
places. The expression in scientific notation is 7.5×10^8.

13. D

Start by writing the first factor in the expression, which is 6.02. Then,
move the decimal point 6 places to the left because of the exponent of
−6. The standard notation is 0.00000602.

14. D

The exponent of $\frac{3}{2}$ indicates the cube of the square root of the base.

In other words, the three in the numerator is a power and the two in the denominator is the root. First find the square root of the base, which is 5. Then, raise 5 to the third power: $5^3 = 5 \times 5 \times 5 = 125$.

15. B

The largest perfect square factor of 48 is 16. Therefore, the radical can be expressed as $\sqrt{16} \times \sqrt{3} = 4\sqrt{3}$.

16. E

Because 81 is a perfect square, simplify inside the parentheses first to get $(9 \times 9)^2 = 81^2 = 6,561$.

17. E

First, simplify to see if there is a common radicand. Since $\sqrt{28} = \sqrt{4} \times \sqrt{7} = 2\sqrt{7}$, now combine the terms by adding the coefficients and keeping the common radicand. $2\sqrt{7} + 3\sqrt{7} = 5\sqrt{7}$.

18. D

Multiply the coefficients by coefficients, and radicals by radicals. The expression becomes $8 \times 4 \times \sqrt{3} \times \sqrt{2} = 32\sqrt{6}$.

19. C

To rationalize the denominator, multiply numerator and denominator

by $\sqrt{7}$. $\frac{11}{\sqrt{7}} \times \frac{\sqrt{7}}{\sqrt{7}} = \frac{11\sqrt{7}}{\sqrt{49}} = \frac{11\sqrt{7}}{7}$.

20. E

Write the numerator and denominator as separate radicals, and then simplify by multiplying by $\sqrt{7}$. $\sqrt{\frac{12}{7}} = \frac{\sqrt{12}}{\sqrt{7}} \times \frac{\sqrt{7}}{\sqrt{7}} = \frac{\sqrt{84}}{\sqrt{49}} = \frac{\sqrt{4} \times \sqrt{21}}{7} = \frac{2\sqrt{21}}{7}$.

CHAPTER 8

Geometry

BUILDING BLOCK QUIZ

Start off your study of geometry with this 10-question Building Block Quiz. The initial questions will help you to review the material from the previous chapter on powers and roots. The succeeding questions are a warm-up for the concepts presented in this chapter. Detailed answer explanations are provided to help you find your errors and clarify concepts.

1. Simplify -6^3
   (A) -18 (B) -216 (C) 18 (D) 216 (E) -63

2. Simplify $2^2 \times 2^3 \times 2^0$
   (A) 32 (B) 0 (C) 64 (D) 10 (E) 12

3. Simplify 3^{-3}
   (A) -9 (B) -27 (C) $-\dfrac{1}{27}$ (D) $\dfrac{1}{9}$ (E) $\dfrac{1}{27}$

4. What is the correct classification, by sides and angles, of the following triangle?

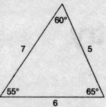

(A) right scalene   (B) obtuse isosceles

(C) right isosceles   (D) acute scalene

(E) acute equilateral

5. What is the measure of the complement of a 47° angle?

(A) 133°   (B) 53°   (C) 43°   (D) 3°   (E) 33°

6. In the figure below, $AB \parallel CD$. What type of angles are angles ∠1 and ∠5?

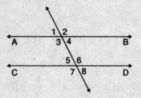

(A) complementary   (B) corresponding   (C) vertical

(D) alternate interior   (E) alternate exterior

7. Which of the following is not true?

(A) All rhombuses are squares.

(B) All squares are rectangles.

(C) All parallelograms are quadrilaterals.

(D) All rectangles are parallelograms.

(E) All squares are rhombuses.

8. Which of the following sets of lengths cannot be the side lengths of a triangle?

   (A) {3, 4, 5}   (B) {4, 4, 5}   (C) {6, 8, 11}

   (D) {9, 13, 21}   (E) {8, 8, 20}

9. Given the figure below, and $\overline{AB} \parallel \overline{CD}$, what is the measure of angle ∠7?

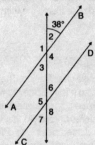

   (A) 22°   (B) 112°   (C) 32°   (D) 38°   (E) 132°

10. Which of the following is always true?

   (A) Vertical angles are congruent.

   (B) Vertical angles are supplementary.

   (C) Vertical angles are complementary.

   (D) All of the above are true.

   (E) None of the above are true.

Answers and Explanations

1.   B

According to order of operations, first evaluate 6 to the third power and then take its opposite. $-(6^3) = -(6 \times 6 \times 6) = -(36 \times 6) = -216$.

2. A

When you multiply numbers with the same base, (in this case, 2), keep the base and add the exponents: $2^2 \times 2^3 \times 2^0 = 2^{2+3+0} = 2^5$. Two to the fifth power is $2 \times 2 \times 2 \times 2 \times 2 = 32$.

3. E

This problem reviews negative exponents. 3^{-3} is not a negative number. The negative exponent directs you to take the reciprocal of the base (in this case, 3) and raise this to the third power: $3^{-3} = \left(\frac{1}{3}\right)^3 = \frac{1^3}{3^3} = \frac{1}{27}$.

4. D

This question covers triangle classification. The triangle has three angles that are all less than 90°, so it is an acute triangle. Each of the three sides has different length, so it is a scalene triangle. The correct classification is acute scalene. Choice (A) would be a triangle with exactly one 90° angle and three sides with different lengths. Choice (B) would be a triangle with exactly one obtuse angle and two congruent sides. Choice (C) would be a triangle with exactly one 90° angle and two congruent sides. Choice (E) would be a triangle with three congruent sides and three angles with measure of 60°.

5. C

The complement of an angle is an angle whose degree measure, when added to the degree measure of the given angle, has a sum of 90°. The complement of 47° is thus 90 − 47 = 43°.

6. B

This problem tests your knowledge of the angle relationships formed when a transversal line cuts through two parallel lines. This pair of angles is on the same side of the transveral line, one is in the interior and one is in the exterior of the parallel lines. The name for this special pair is corresponding angles.

7. A

This problem touches upon some of the facts concerning quadrilaterals. All rhombuses are not necessarily squares. The requirement for a rhombus is that it is a parallelogram with four equal side lengths. The only rhombuses that are squares are those that have four right angles.

8.   E

This question focuses on the requirements for the lengths of the sides of a triangle. The side lengths of a triangle must pass the test that the sum of any two sides must be greater than the third side. For choice (E), $8 + 8 = 16$, which is less than 20.

9.   D

This problem tests your knowledge of the angle relationships formed when a transversal line cuts through two parallel lines. $\angle 7$ has the same degree measure as $\angle 2$, since these two angles are alternate exterior angles. The measure of $\angle 7 = 38°$ because the measure of $\angle 2 = 38°$.

10.   A

Vertical angles are always congruent. Vertical angles would only be complementary if their measures were each $45°$, to make a sum of $90°$. Vertical angles would only be supplementary if their measures were each $90°$, to make a sum of $180°$.

BASIC GEOMETRY ELEMENTS

Start off your study of geometry by learning some basic terms and definitions.

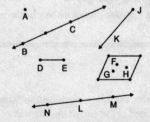

Point – a position in space. A point has no length or width. An example is point A.

Line – The shortest distance between two points determines a line. A line has no width, and an infinite length. An example is line $\overrightarrow{BC}$.

Line segment – a piece of a line. A segment includes two points, called endpoints, and all the points in between. An example is $\overline{DE}$.

Plane – Any three non-collinear points determine a plane. A plane has an infinite width and length. An example is plane FGH.

Ray – one half of a line. A ray includes an endpoint and all the collinear points on one side of this endpoint. An example is $\overrightarrow{JK}$, or $\overrightarrow{LM}$ (one half of line $\overleftrightarrow{NM}$).

When naming lines, line segments, or planes, the order of the points does not matter. In the above definitions, segment $\overline{DE}$ could also be named $\overline{ED}$; line $\overleftrightarrow{BC}$ could also be named $\overleftrightarrow{CB}$, and plane FGH could also be named GFH, or other variations. But when naming a ray, the endpoint must be named first. In the above figure, there is a ray $\overrightarrow{LM}$ and a different ray $\overrightarrow{LN}$. Both have the same starting point of L, but they contain all other points that are different on the opposing sides of the line $\overleftrightarrow{NM}$. As a rule, line segments, lines, planes, and rays all have an infinite number of points.

MATH SPEAK

Two geometric figures are **congruent** if each has the same measure. This relationship can apply to several geometric figures. The symbol for congruence is ≅.

A segment **bisector** is a line or line segment that divides the segment into two congruent segments.

The **midpoint** of a segment is where the bisector intersects the segment.

Practice 1

Answer true or false for the following questions. Answers and explanations are located at the end of the chapter.

1. **T**   **F**   When naming a line segment, the order of the points named does not matter.

2. **T**   **F**   When naming a ray, the order of the points named does not matter.

3. **T**   **F**   A line segment has an infinite number of points.

4. **T**   **F**   If the length of segment $\overline{BC}$ is 14 cm, and B is the midpoint of segment $\overline{AC}$, then the measure of segment $\overline{AC}$ is 7 cm.

Special Lines

There are three kinds of special lines:

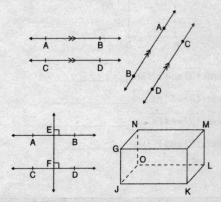

Parallel lines – lines in the same plane that do not intersect. In the figure above, lines $\overleftrightarrow{AB}$ and $\overleftrightarrow{CD}$ are parallel as shown by $AB \parallel CD$. Line segments, rays, and planes can also be parallel.

Perpendicular lines – lines in the same plane that intersect at one point and form four 90° angles. In the figure above, line $\overleftrightarrow{EF}$ is perpendicular to both of the lines $\overleftrightarrow{AB}$ and $\overleftrightarrow{CD}$. This is shown by the symbols $EF \perp AB$ and $EF \perp CD$. Line segments, rays, and planes can also be perpendicular.

Skew lines – lines in different planes that do not intersect. Skew lines exist in three dimensions. In the rectangular solid in the figure above, segments NM and KL are skew.

Practice 2

Use the following figure and the word bank to fill in the blanks.

parallel   perpendicular   skew

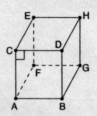

5. Segments $\overline{CD}$ and $\overline{EH}$ are _____.

6. Segments $\overline{CD}$ and $\overline{AC}$ are _____.

7. Segments $\overline{CD}$ and $\overline{BG}$ are _____.

8. Segments $\overline{CD}$ and $\overline{FG}$ are _____.

Angles

> **MATH SPEAK**
>
> An **angle** is defined by two distinct rays that have the same endpoint.

The common endpoint the rays share is called the *vertex* of the angle. When naming an angle, you must use the symbol for angle (∠) and you must name the vertex of the angle in the middle of the three letters. In the following figure, there are several angles, which are numbered for clarity.

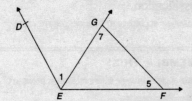

∠1 can be named ∠DEG or ∠GED. ∠5 can be named ∠EFG or ∠GFE. ∠7 can be named ∠FGE or ∠EGF. Note that ∠5 and ∠7 use the same three points, but name different angles, based on the vertex point.

MATH SPEAK

An **angle bisector** is a ray in the interior of the angle that divides the angle into congruent angles with the same measure.

In the figure below, ray $\overrightarrow{BD}$ is an angle bisector: ∠ABD ≅ ∠DBC. Note how congruent angles are indicated on the figure.

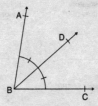

Angle Classification

Angles are classified according to their degree measure.

> ### REMEMBER THIS!
>
> An **acute** angle is an angle whose measure is between 0°
> and 90°.
>
> A **right** angle measures exactly 90°. As shown in the figure
> below, right angles are designated by a small box in the interior
> at the vertex.
>
> An **obtuse** angle is an angle whose measure is between 90°
> and 180°.
>
> A **straight** angle measures exactly 180°.

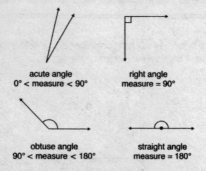

acute angle
0° < measure < 90°

right angle
measure = 90°

obtuse angle
90° < measure < 180°

straight angle
measure = 180°

Special Angle Pair Relationships

Complementary angles are any two angles whose combined
measures equal 90°. It is not necessary for these angles to share
a side.

Supplementary angles are any two angles whose combined measures
equal 180°. It is not necessary for these angles to share a side.

A **linear pair** is two supplementary angles that share a common side and no common interior points. They will form a straight angle, or a line.

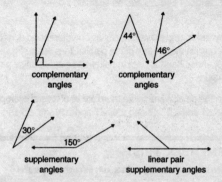

complementary
angles

complementary
angles

supplementary
angles

linear pair
supplementary angles

When a line, called a *transversal*, cuts through two parallel lines, eight angles are formed. These angles are numbered 1 through 8 in the figure below for clarity:

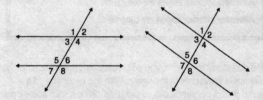

From this condition, several types of angles are defined:

Vertical angles are formed when two lines intersect. They are two angles that share a common vertex but share no common sides. In the figure above, ∠1 and ∠4, ∠2 and ∠3, ∠5 and ∠8, and ∠6 and ∠7 are all examples of vertical angle pairs.

Corresponding angles are two angles on the same side of the transversal, but one is in the interior of the parallel lines and one is in the exterior of the parallel lines. In the figure above, ∠1 and ∠5, ∠2 and ∠6, ∠3 and ∠7, and ∠4 and ∠8 are all examples of corresponding angle pairs.

Alternate interior angles are two angles on different sides of the transversal, both in the interior of the parallel lines. In the figure above, ∠3 and ∠6, and ∠4 and ∠5 are the alternate interior angle pairs in the above figure.

Alternate exterior angles are two angles on different sides of the transversal, both in the exterior of the parallel lines. In the figure above, ∠1 and ∠8, and ∠2 and ∠7 are the alternate exterior angle pairs in the above figure.

Note that in the parallel line figure there are also several examples of linear pairs: ∠1 and ∠2, ∠2 and ∠4, ∠7 and ∠8, and ∠5 and ∠7.

REMEMBER THIS!

When a line, called a transversal, cuts through two parallel lines:

Vertical angles are congruent.

Corresponding angles are congruent.

Alternate interior angles are congruent.

Alternate exterior angles are congruent.

In fact, if you study the figure, you will realize that there are actually only two angle measures in the figure. With the exception of a transversal that is perpendicular to the parallel lines, four acute angles and four obtuse angles are formed. All of the acute angles are congruent and all of the obtuse angles are congruent. In addition, any one of the acute angles paired with any one of the obtuse angles forms a supplementary pair.

Practice 3

Using the figure below, fill in the blanks with the name of the special pair of angles. Answers and explanations are located at the end of the chapter.

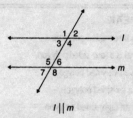

$l \parallel m$

9. Angles ∠5 and ∠8 are called _____ angles.

10. Angles ∠8 and ∠6 are _____ angles.

11. Angles ∠7 and ∠3 are called _____ angles.

12. Angles ∠4 and ∠6 are _____ angles.

13. Angles ∠4 and ∠5 are called _____ angles.

14. Angles ∠7 and ∠2 are called _____ angles.

15. Angles ∠1 and ∠2 are _____ angles.

Using the figure above, if the measure of angle ∠1 = 132°, give the following angle measures:

16. the measure of ∠5

17. the measure of ∠6

18. the measure of ∠7

19. the measure of ∠8

POLYGONS

Polygons are closed geometric figures made up of line segments and angles. Polygons are named based on the number of sides.

> **MATH SPEAK**
>
> A **triangle** is a three-sided polygon.
>
> A **quadrilateral** is a four-sided polygon.
>
> A **pentagon** is a five-sided polygon.
>
> A **hexagon** is a six-sided polygon.
>
> An **octagon** is an eight-sided polygon.

Triangles

The triangle is the most common geometric figure used in mathematics. A triangle is a closed geometric figure that has three sides and three angles. Triangles are classified according to their angles and their side lengths.

> **REMEMBER THIS!**
>
> Triangles are classified according to their angles as follows:
>
> An **acute** triangle has three acute angles.
>
> An **obtuse** triangle has exactly one obtuse angle and two acute angles.
>
> A **right** triangle has exactly one right angle, and two acute angles.
>
> Triangles are classified according to their sides as follows:
>
> A **scalene** triangle has three sides of different length.
>
> An **isosceles** triangle has two sides of the same length. This triangle also has two angles with the same measure – the angles opposite to the sides of the same length.
>
> An **equilateral** triangle has all three sides of the same length. This triangle also has three congruent angles, each with measure of 60°.

Thus, every triangle can be classified in two ways, either by angle or by sides. The figure below gives some examples of triangles and their classifications.

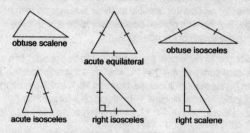

obtuse scalene

acute equilateral

obtuse isosceles

acute isosceles

right isosceles

right scalene

The Angles of a Triangle

There are some important facts about triangles that are important to understand when performing operations with angles.

> **REMEMBER THIS!**
>
> The sum of the measures of the angles in a triangle is 180°.
>
> The measure of an exterior angle to a triangle is equal to the sum of the two remote interior angles of the triangle.
>
> In a right triangle, the two acute angles are complementary, that is, the sum of the measure of these angles is equal to 90°.

For example, in the figure below, $\triangle ABC$ is a right triangle. Therefore, $\angle BAC$ and $\angle ACB$ are complementary. The measure of $\angle ACB = 90 - 44 = 46°$. In addition, the measure of $\angle ACD$, which is an exterior angle to triangle $\triangle ABC$, is equal to the sum of $\angle CAB$ and $\angle ABC$, the two remote interior angles. The measure of $\angle ACD = 90 + 44 = 134°$. You may also notice that $\angle ACD$ and $\angle BCA$ form a linear pair, and therefore their angle measures sum to 180°.

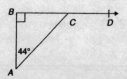

The Sides of a Triangle

The sum of the lengths of any two sides of a triangle must be larger than the length of the third side. You may be given three side lengths, and then be asked if it is possible for these side lengths to form a triangle. Test the sum of each possible pair to be sure that this sum exceeds the length of the other side.

For example, to determine if 4, 7, and 12 can be the sides of a triangle, check each sum pair: $4 + 7 = 11$, which is not greater than 12. This set cannot be the sides of a triangle. However, 4, 4, and 7 can be the sides of a triangle. Test the sides: $4 + 4 = 8$, which is bigger than 7; the other two tests would be $7 + 4 = 11$, which is bigger than 4.

If you are given two side lengths of a triangle, the third side must be smaller than the sum of the two given sides, and larger than the difference of the two sides. For example, if two sides of a triangle are given as 14 cm and 21 cm, the other side must be between $(21 - 14)$ and $(14 + 21)$. Using x to represent the length of the third side, $7 < x < 35$.

Practice 4

Repetition is the key to mastery! Try these problems concerning triangles. Answers and explanations are located at the end of the chapter.

20. A triangle has angles that measure 85°, 50°, and 45°, and sides of length 6 inches, 7 inches, and 10 inches. What is the classification of this triangle?

21. The measure of two of the angles in a triangle are 35° and 35°. What is the measure of the third angle?

22. Given the triangle below, with angle measures as shown, what is the measure of angle ∠BCD?

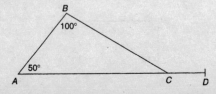

23. Two sides of a triangle measure 3 cm and 7 cm. What are the possible measures of the third side?

Quadrilaterals

Quadrilaterals are four-sided polygons. There are two major classifications of common quadrilaterals:

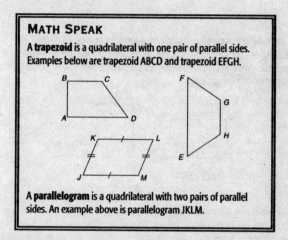

MATH SPEAK

A **trapezoid** is a quadrilateral with one pair of parallel sides. Examples below are trapezoid ABCD and trapezoid EFGH.

A **parallelogram** is a quadrilateral with two pairs of parallel sides. An example above is parallelogram JKLM.

For every quadrilateral, the sum of the measure of the angles is 360°. In a parallelogram, both pairs of opposite sides are parallel and congruent. Opposite angles are also congruent.

Trapezoids and parallelograms can be further classified into subcategories.

An **isosceles trapezoid** is a trapezoid with non-parallel sides that are equal in length. Examples are trapezoids PQRS and WXYZ, as shown on the next page. Because the non-parallel sides are congruent, there are two pairs of congruent angles. In addition, an isosceles trapezoid has *congruent* diagonals.

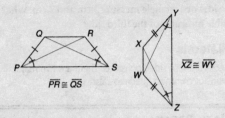

$\overline{PR} \cong \overline{QS}$

$\overline{XZ} \cong \overline{WY}$

A **rectangle** is a parallelogram with four right angles, as shown in rectangles LMNO and STUV. Like all parallelograms, the rectangle has opposite sides and opposite angles that are congruent. In addition, the diagonals are *congruent*.

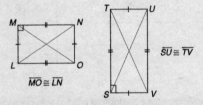

$\overline{MO} \cong \overline{LN}$

$\overline{SU} \cong \overline{TV}$

A **rhombus** is a parallelogram with four congruent sides, as shown in rhombus ABCD and WXYZ below. Like all parallelograms, the rhombus has opposite sides and opposite angles that are congruent. In addition, the diagonals are *perpendicular*.

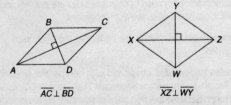

$\overline{AC} \perp \overline{BD}$

$\overline{XZ} \perp \overline{WY}$

A **square** is both a rhombus and a rectangle. Therefore it is a parallelogram with four 90° angles and all sides are congruent. In addition, the diagonals of a square are both *congruent* and *perpendicular*. Square JKLM is shown on the following page.

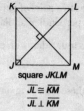

square *JKLM*

$$\overline{JL} \cong \overline{KM}$$

$$\overline{JL} \perp \overline{KM}$$

The tree below summarizes the relationship between the different classifications of special quadrilaterals. The characteristics are listed under each heading. All polygons lower in the tree share the characteristics of the polygons above it in the tree.

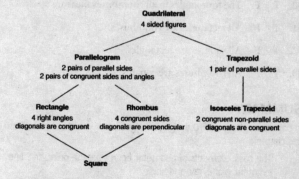

Practice 5

For the problems below, match the description with the most specific classification from the word bank. Answers and explanations are located at the end of the chapter.

parallelogram   trapezoid   rectangle   square   rhombus

24. This quadrilateral has four right angles, four congruent sides and diagonals that are perpendicular and congruent.

25. This quadrilateral has two pairs of parallel sides.

26. This quadrilateral has four right angles and congruent diagonals.

27. This quadrilateral has two pairs of parallel sides, four congruent sides and perpendicular diagonals.

28. This quadrilateral has four sides, with exactly one pair of parallel sides.

Answer true or false for these following statements about quadrilaterals.

29. **T F** If one of the angles in a parallelogram measures 60°, the measures of the other angles are 60°, 40°, and 40°.

30. **T F** The four angles in all rhombuses measure 90°.

31. **T F** All rectangles are rhombuses.

32. **T F** All squares are rectangles.

33. **T F** All rhombuses are squares.

SUMMARY

In conclusion, here are 10 things about geometry you should take from this chapter:

- The basic elements of geometric figures are the point, line, line segment, plane, ray, and angle.
- Angles are classified as acute, obtuse, right, or straight.
- Complementary angles sum to 90°; supplementary angles sum to 180°.
- Special congruent angle pairs are formed when a transversal line cuts two parallel lines.
- These pairs are: corresponding, vertical, alternate interior, and alternate exterior angles.
- Polygons are named according to the number of sides.
- Triangles are classified in two ways, by sides and by angles.
- The sum of the degree measures of the angles in a triangle is 180°.

- The sum of the measures of any two sides of a triangle must be larger than the measure of the third side.
- Quadrilaterals are classified into two major groups: parallelograms and trapezoids.
- Parallelograms are classified into rectangles and rhombuses; the square is both of these.
- The sum of the measure of the angles in a quadrilateral is 360°.

Practice Answers and Explanations

1. **True**

2. **False**

When you name a ray, the endpoint must be named first.

3. **True**

4. **False**

You are given that B is the midpoint of segment $\overline{AC}$. This means that segment $\overline{AC}$ is twice the length of either segment $\overline{AB}$ or $\overline{BC}$. You are told that the measure of segment $\overline{BC} = 14$ cm, so the measure of $\overline{AC} = 2 \times 14 = 28$ cm.

5. **parallel**

These segments are in the same plane, (the top face), and do not intersect. Perpendicular segments intersect to form right angles. Skew segments are segments in different planes that do not intersect.

6. **perpendicular**

These segments intersect to form right angles. Parallel segments are segments in the same plane that do not intersect. Segments that are skew are segments in different planes that do not intersect.

7. **skew**

These segments are in different planes, and they do not intersect. They would not intersect even if they were extended to be lines. Parallel segments are in the same plane and do not intersect. Perpendicular segments intersect to form right angles.

8. parallel

These line segments are parallel in the plane that cuts diagonally through the solid.

9. vertical

Vertical angles are formed when two lines intersect. They share a common vertex, but are non-adjacent.

10. supplementary

Supplementary angles are a pair of angles whose combined measure is 180°. This angles are linear and are always supplementary, because together they form a straight line whose measure is 180°.

11. corresponding

These angles are congruent angles on the same side of the transversal; one is in the interior and the other is in the exterior of the parallel lines. This, by definition, is a pair of corresponding angles.

12. supplementary

When two parallel lines are cut by a transversal line, any one of the acute angles and any one of the obtuse angles form a supplementary pair. In the figure, $\angle 4$ is an obtuse angle, and $\angle 6$ is an acute angle.

13. alternate interior

These angles are both in the interior of the parallel lines, and are on opposite sides of the transversal line. This pair of congruent angles are, by definition, named alternate interior angles.

14. alternate exterior

These angles are both in the exterior of the parallel lines, and they are on opposite sides of the transversal line. This pair of congruent angles are by definition alternate exterior angles.

15. supplementary

Supplementary angles are a pair of angles whose combined measure is 180°. These angles are linear and are always supplementary, because together they form a straight line whose measure is 180°.

16.   132°

Angles ∠5 and ∠1 are corresponding angles; they are congruent, thus their measures are equal.

17.   48°

Angle ∠6 and ∠5 are supplementary; they form a linear pair. 180 − 132 = 48°.

18.   48°

∠1 and ∠8 are alternate exterior angles and therefore congruent. ∠8 and ∠7 are supplementary.

19.   132°

One relationship you can use is that angles ∠5 and ∠8 are vertical angles. They are congruent and thus have the same measure.

20.   acute scalene

All of the angles in this triangle measure less than 90°, so it is an acute triangle. None of the sides have equal measure, so the triangle is scalene. The classification is acute scalene.

21.   110°

The sum of the angles in a triangle is equal to 180°. 180 − 35 − 35 = 110°.

22.   150°

The measure of the exterior angle ∠BCD is equal to the sum of the two remote interior angles: 100 + 50 = 150°.

23.   Greater than 4 cm and less than 10 cm

The third side is between the sum and the difference of the two given side measures. If x represents the third side, then $(7 − 3) < x < (7 + 4)$, or $4 < x < 10$.

24.   square

25.   parallelogram

26.   rectangle

27.   rhombus

28. trapezoid

29. False

In a parallelogram, there are two pairs of congruent angles. Therefore, if one of the angles measures 60°, then another angle measures 60°. The sum of all the angles is 360°, so subtract to find the measure of the other pair. $360 - 60 - 60 = 240°$. Thus, the measure of each of the other two congruent angles is 120°.

30. False

The only type of rhombus with 90° angles is the square.

31. False

The only rectangle that is a rhombus is the square.

32. True

33. False

Some, but not all, rhombuses are squares.

CHAPTER 8 TEST

You have now reviewed the key concepts of geometry and have tried various types of problems. Repetition is the key to mastery of any math concept! Use the chapter material, including the practice questions throughout, to assist you in solving these problems. The answer explanations that follow will provide additional help.

Use the following figure for questions 1 and 2.

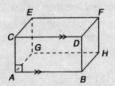

1. Which of the following is true:

    (A) $\overline{AB} \parallel \overline{GH}$ (B) $\overline{AB} \perp \overline{AC}$ (C) $\overline{AB} \parallel \overline{EF}$

    (D) all of the above   (E) none of the above

2. Which of the following segment pairs are skew?
   (A) AB and EF   (B) BH and CD
   (C) CA and FH   (D) All of the segment pairs are skew.
   (E) None of the segment pairs are skew.

3. If angle ∠ABC measures 85°, then it is
   (A) acute   (B) obtuse   (C) straight
   (D) right   (E) isosceles

4. *In the figure below, $\overline{AB}$ is the bisector of $\overline{CD}$.*

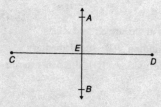

   If $\overline{CE}$ = 14 cm, which of the following is true?
   (A) $\overline{AE}$ = 14 cm   (B) $\overline{AE} \cong \overline{EB}$ (C) $\overline{ED}$ = 14 cm
   (D) $\overline{CD}$ = 30 cm   (E) all of the above

5. Which of the following is true of the figure below?

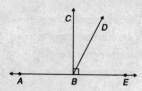

   (A) ∠CBA and ∠CBE are complementary
   (B) ∠CBD and ∠DBE are supplementary
   (C) ∠ABD and ∠DBE form a linear pair
   (D) All of the above are true.
   (E) None of the above are true.

6. Which of the angle measures is supplementary to an angle of 19°?

   (A) 19° (B) 81° (C) 71° (D) 31° (E) 161°

Use the following figure for questions 7, 8, 9, and 10.

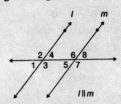

7. What type of angles are ∠1 and ∠8?

   (A) corresponding (B) supplementary

   (C) alternate interior (D) alternate exterior

   (E) vertical

8. If the measure of ∠2 = 132°, then the measure of ∠6 =

   (A) 68° (B) 132° (C) 48°

   (D) 32° (E) none of the above

9. Which of the following is NOT true?

   (A) ∠2 and ∠3 are vertical angles

   (B) ∠4 and ∠7 are supplementary angles

   (C) ∠5 and ∠1 are congruent angles

   (D) ∠3 and ∠5 are congruent angles

   (E) ∠4 and ∠5 are alternate interior angles

10. What angle is supplementary to ∠1?

   (A) ∠7 (B) ∠5 (C) ∠8

   (D) ∠4 (E) none of the above

11. Classify the triangle below:

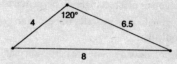

(A) acute isosceles   (B) obtuse isosceles
(C) obtuse scalene   (D) acute scalene
(E) acute obtuse

12. If two angles of a triangle measure 68° and 110°, what is the measure of the third angle?

(A) 12°   (B) 2°   (C) 22°   (D) 92°   (E) 192°

13. Given the following figure, what is the measure of angle ∠CDF?

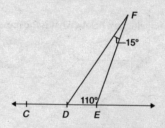

(A) 100°   (B) 85°   (C) 75°   (D) 65°   (E) 125°

14. One acute angle of a right triangle measures 37°. What is the measure of the other acute angle?

(A) 143°   (B) 37°   (C) 63°   (D) 53°   (E) 43°

15. If two sides of a triangle have measures of 2 cm and 7 cm, what is the possible values for the third side, in cm?

(A) $7 < x < 2$ (B) $2 < x < 7$ (C) $1 < x < 8$
(D) $5 < x < 9$ (E) $9 < x < 5$

16. Which of the following is ALWAYS true?

   (A) The diagonals of a rectangle are perpendicular.
   (B) The diagonals of an isosceles trapezoid are perpendicular.
   (C) The diagonals of a square are perpendicular.
   (D) All of the above are true.
   (E) None of the above are true.

17. Which of the following is true?

   (A) All rectangles are squares.
   (B) All squares are parallelograms.
   (C) All rhombuses are trapezoids.
   (D) All rhombuses have four congruent angles.
   (E) All parallelograms are rectangles.

18. Given the parallelogram below, what is the measure of ∠BCD?

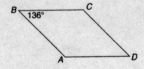

   (A) 136°
   (B) 44°
   (C) 124°
   (D) 64°
   (E) Cannot be determined from the information given.

19. In isosceles trapezoid JKLM below, what is the measure of ∠KJM?

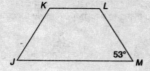

(A) 53°

(B) 127°

(C) 147°

(D) 47°

(E) Cannot be determined from the information given.

20. Given the rectangle ABCD below, and $\overline{AB}$ = 3 cm, $\overline{BC}$ = 4 cm and $\overline{AC}$ = 5 cm, what is the measure of $\overline{BD}$?

(A) 3 cm

(B) 4 cm

(C) 5 cm

(D) 7 cm

(E) Cannot be determined from the information given.

Answers and Explanations

1. **D**

All of the statements are true. Segments $\overleftrightarrow{AB}$ and $\overleftrightarrow{GH}$ are parallel; they are opposite sides of the rectangular base. Segments $\overline{AB}$ and $\overline{AC}$ are perpendicular, as shown by the right angle designation. Segments $\overleftrightarrow{AB}$ and $\overleftrightarrow{EF}$ are parallel; they are in a plane that runs diagonally through the solid.

2. B

Both of the segment pairs, $\overleftrightarrow{AB} - \overleftrightarrow{EF}$ and $\overleftrightarrow{CA} - \overleftrightarrow{FH}$, are parallel through a plane that runs diagonally through the solid. The only skew pair is $\overleftrightarrow{BH}$ and $\overleftrightarrow{CD}$. There is no plane in which they reside together.

3. A

An angle whose measure is between 0° and 90° is an acute angle.

4. C

It is given that line $\overleftrightarrow{AB}$ is the bisector of segment $\overline{CD}$, thus it divides segment CD into two congruent segments. The given statement tells nothing about the segments $\overline{AE}$ or $\overline{EB}$; choices (A) and (B) are thus not necessarily true. Segment $\overleftrightarrow{CD}$ is 28 cm long, not 30 cm long, as stated in choice (D).

5. C

The only true statement is choice (C), that ∠ABD and ∠DBE form a linear pair; they are adjacent and together form a straight angle of 180°. ∠CBA and ∠CBE are supplementary, not complementary as stated in choice (A). ∠CBD and ∠DBE are complementary, not supplementary as stated in choice (B).

6. E

The measure of an angle and its supplement have a sum of 180°. So the measure of the supplement to a 19° angle is 180 − 19 = 161°.

7. D

These angles are alternate exterior angles. They are both in the exterior of the parallel lines, on opposite sides of the transversal line.

8. B

Angles ∠2 and ∠6 are corresponding angles, therefore their angle measures are the same.

9. D

All statements are true, except choice (D) which states that ∠3 and ∠5 are congruent. Angles ∠4 and ∠7 are supplementary, because ∠4 and ∠3 are a linear pair, and ∠3 is congruent to ∠7 because they are corresponding angles. Angles ∠5 and ∠1 are congruent because they are corresponding angles.

10. A

The angles that are supplementary to ∠1, or add up to equal 180° are angles ∠2, ∠3, ∠6, or ∠7. All of the other angles are congruent to ∠1.

11. C

There is one angle with measure greater than 90°. Therefore, the triangle is obtuse. All of the sides have different measures, so the triangle is scalene. The classification is obtuse scalene. An acute isosceles would have three angles with measures less than 90°, and two congruent sides. A triangle classified as obtuse isosceles has two congruent sides. In an acute scalene triangle, all three angles must have measures less than 90°, which is not true of this triangle. Choice (E) is also incorrect, because it gives two different classifications based on the angles of a triangle. The given triangle is an obtuse, not an acute triangle.

12. B

The sum of the measure of the angles in a triangle is 180°. The third angle measure must be $180 - 110 - 68 = 2°$.

13. E

Angle ∠CDF is an exterior angle; its measure is equal to the sum of the two other remote interior angles, 110° and 15°. The measure of ∠CDF $= 110 + 15 = 125°$.

14. D

The two acute angles in a right triangle are complementary; their measures have a sum of 90°. Thus, the measure of the other acute angle is $90 - 37 = 53°$.

15. D

The possible values for the third side are between the difference, $(7-2 = 5)$ and the sum $(7 + 2 = 9)$. So the third side is between 5 and 9.

16. C

In choice (A), the diagonals of a rectangle are not always perpendicular; this is only true for a square. In choice (B), the diagonals of an isosceles trapezoid are always congruent, not necessarily perpendicular. All squares are always rhombuses. Therefore, the diagonals of a square are always perpendicular.

17. B

The only true statement is that all squares are parallelograms; they have two sets of parallel and congruent sides. Choice (A) is not true. Only some rectangles are squares. Choice (C) is never true. No rhombuses are trapezoids. Choice (D) is only true when the rhombus is a square. Choice (E) is only sometimes true.

18. B

The measure of the four angles in a parallelogram add to total 360°. There are two sets of congruent angles. Therefore, 360 − 136 − 136 = 88°, which is the sum of the second pair of congruent angles. Thus, each of these angles, one of which is ∠BCD, is 44°.

19. A

In an isosceles trapezoid, the base angles are congruent. Angle ∠LMJ and ∠KJM are the base angles and thus have the same measure of 53°.

20. C

In every rectangle, the diagonals are congruent. Since the diagonal segment $\overline{AC}$ = 5 cm, the other diagonal segment $\overline{BD}$ is also 5 cm.

CHAPTER 9

Measurement

Measurement is an important aspect of geometry. To prepare for your study, try the following Building Block Quiz. Some of the questions will review last chapter's material on geometry. Score yourself and read over the answer explanations to prepare for the material in this chapter.

BUILDING BLOCK QUIZ

1. What is the measure of the supplement to an angle of measure 36°?

   (A) 64° (B) 44° (C) 54° (D) 144° (E) 36°

2. One acute angle of a right triangle measures 19°. What is the measure of the other acute angle?

   (A) 90° (B) 161° (C) 19° (D) 81° (E) 71°

3. Which of the following statements about quadrilaterals is true?

   (A) All rhombuses are squares.

   (B) All parallelograms have diagonals that are congruent.

   (C) All trapezoids have exactly one set of parallel sides.

   (D) All parallelograms have exactly one set of parallel sides.

   (E) The sum of the measure of the angles in a quadrilateral is 180°.

4. Find the perimeter of the following polygon:

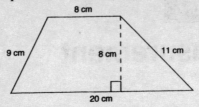

(A) 56 cm   (B) 180 cm^2 (C) 48 cm

(D) 160 cm^2 (E) 39 cm

5. Lisa wants to tile her kitchen floor. The floor is a rectangle with dimensions of 14'6" by 12'. How many square feet of tiles are needed to cover the floor?

(A) 174 ft^2 (B) 175.2 ft^2 (C) 26.6 ft^2

(D) 53.5 ft^2 (E) 1,752 ft^2

6. What is the area of a circle with a 24-inch diameter? Use 3.14 for π.

(A) 1,808.64 in^2 (B) 576 in^2 (C) 75.36 in^2

(D) 452.16 in^2 (E) 144 in^2

7. What is the surface area of the following rectangular solid?

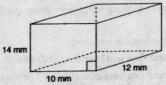

(A) 840 mm^2 (B) 1,680 mm^2 (C) 1,008 mm^2

(D) 428 mm^2 (E) 856 mm^2

8. How much water will a rectangular swimming pool with dimensions of 60' × 24' × 6' hold?

(A) 8,640 ft³   (B) 90 ft³   (C) 1,440 ft³

(D) 540 ft³   (E) 2,880 ft³

9. What is the length of the diagonal of a square with sides of 25 cm?

(A) 62.5 cm   (B) 50 cm   (C) 10 cm

(D) 35.36 cm   (E) 25 cm

10. In the following right triangle, △ABC, what is the tangent of angle ∠BAC?

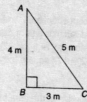

(A) 5   (B) 0.75   (C) 0.6   (D) $\frac{4}{3}$ (E) $\frac{4}{5}$

Answers and Explanations

1.   **D**

The degree measure of an angle and its supplement have a sum of 180°. The degree measure of the supplement to an angle of 36° is thus 180 − 36 = 144°.

2.   **E**

This question reviews the sum of the degrees in a triangle and the properties of right triangles. The two acute angles in a right triangle add up to 90°. Therefore, the angle has measure of 90 − 19 = 71°.

3. C

This problem reviews basic facts about quadrilaterals. The only true statement is that a trapezoid has exactly one set of parallel sides. Statement (A) is false; only some rhombuses are squares. Statement (B) is false; only the rectangle has diagonals that are congruent. Statement (D) is false; parallelograms have two sets of parallel sides. Statement (E) is false; the sum of the measure of the angles in a quadrilateral is 360°.

4. C

This question reviews the concept of perimeter. To find the perimeter, add up the measures of the sides of the trapezoid. The height of 8 cm is not used to find the perimeter: $9 + 8 + 11 + 20 = 48$ cm.

5. A

This question reviews the concept of area. You need to find the area when something is covered, such as a floor. Six inches is half of one foot, so change the dimensions to 14.5′ by 12′. To find the square footage, multiply: $14.5 \times 12 = 174$ square feet.

6. D

Question 6 tests your knowledge of the area of a circle. The formula to use is $A = \pi r^2$, where r is the radius, and 3.14 is used for π. The radius is one-half of the 24 inch diameter, or 12 inches. $A = 3.14 \times 12^2 = 3.14 \times 144 = 452.16$ in^2.

7. E

This question reviews surface area. To find surface area, find the area of each face and add them all together. This solid has six faces, with opposite faces that are congruent. Each face is a rectangle, so the area of each face is *base* × *height*. So the surface area is *front* + *back* + *top* + *bottom* + *left* + *right*. Surface area = $(10 \times 14) + (10 \times 14) + (10 \times 12) + (10 \times 12) + (14 \times 12) + (14 \times 12)$. Because opposite faces are congruent, this simplies to $(2 \times 140) + (2 \times 120) + (2 \times 168)$, or $280 + 240 + 336 = 856$ mm^2.

8. A

This question tests your knowledge of volume. The volume of a rectangular solid is found by multiplying *length* × *width* × *height*. Therefore the volume is $60 \times 24 \times 6 = 8,640$ ft^3.

9. D

Question 9 reviews an application of the Pythagorean theorem. Together with any two adjacent sides, the diagonal forms the hypotenuse of a right triangle. Use the Pythagorean theorem, $c^2 = a^2 + b^2$, to solve for the hypotenuse, c. $c^2 = 25^2 \times 25^2$, or $c^2 = 625 + 625$. $c^2 = 1{,}250$. To now solve for c, take the square root of 1,250 to get approximately 35.36 cm.

10. B

This problem tests your knowledge of trigonometry. The tangent is the ratio of the $\dfrac{\text{the opposite side}}{\text{the adjacent side}}$ to the angle in question. In this figure, the opposite side is 3 m and the adjacent side is 4 m. The tangent is thus $\dfrac{3}{4} = 0.75$.

MEASUREMENT

This chapter will explore the concepts of perimeter, area, surface area, and volume. In addition, you will review two important applications of right triangle measurement: the Pythagorean theorem and trigonometry.

Perimeter

> **MATH SPEAK**
>
> The **perimeter** is the distance around a polygon.

Some real world applications of perimeter are fencing a yard or framing a picture. To find the perimeter of a polygon, add together the lengths of all its sides. For example, the perimeter of the following trapezoid is $21 + 8 + 5 + 12 = 46$ cm.

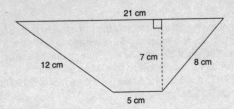

Notice that the height of the trapezoid (7 cm) is indicated on the figure. However, this fact is not needed to calculate the perimeter.

FLASHBACK

Recall these facts from chapter 8 on geometry:

Squares have four congruent sides.

Parallelograms have two pairs of congruent sides.

A **rectangle** is a parallelogram.

These facts will help you to figure out the formulas for the perimeter of a square and a parallelogram.

REMEMBER THIS!

The perimeter of a square with side x is $P = 4x$.

The perimeter of a parallelogram with sides y and z is $P = 2y + 2z$.

Practice 1

Repetition is the key to mastery! Try these questions concerning perimeter. Answers and explanations are located at the end of the chapter.

1. Find the perimeter of the following triangle:

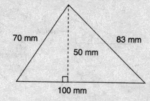

2. What is the perimeter of a square with side lengths of 16 m?

3. Find the perimeter of the following parallelogram:

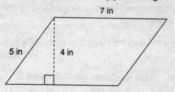

Area

> **MATH SPEAK**
>
> The **area** of a polygon is the amount of square units needed to cover the polygon.

Some real world applications of area include carpeting or tiling a floor and painting or wallpapering walls. Area is calculated by multiplying two entities, so the units are always in square units. To use the formulas to calculate area, you need to know how to recognize and find the height of a polygon.

MATH SPEAK

The **height** of a polygon is the length of the segment that is perpendicular to a side of the polygon (called the base). Study the figure below to recognize the height of different polygons.

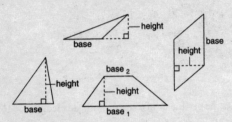

To distinguish the height from a side of the polygon, it is usually shown as a dotted line. Sometimes, as in rectangles and squares, the height is also one of the sides.

The formulas used to find the area of common polygons are as shown below:

Area of Common Polygons

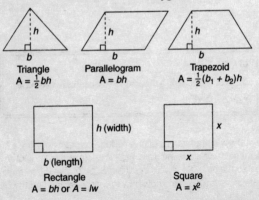

Triangle
$A = \frac{1}{2}bh$

Parallelogram
$A = bh$

Trapezoid
$A = \frac{1}{2}(b_1 + b_2)h$

Rectangle
$A = bh$ or $A = lw$

Square
$A = x^2$

To find the area of a polygon, substitute the given lengths into the formula and simplify. The units for area are always square units.

Use the following figure of a triangle and a parallelogram to find the respective areas:

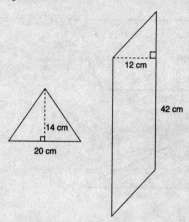

To find the area of the triangle shown above, substitute in 14 m for the height (the length of the segment perpendicular to the base), and substitute in 20 m for the base. Use the formula for the area of a triangle, that is $A = \frac{1}{2} bh$. The area is $\frac{1}{2} \times 20 \text{ cm} \times 14 \text{ cm} = 140 \text{ cm}^2$.

To find the area of the parallelogram, use the formula $A = bh$, that is $A = 42 \text{ cm} \times 12 \text{ cm} = 504 \text{ cm}^2$.

Practice 2

Fill in the blanks. Answers are located at the end of the chapter.

4. Area is measured in _____ units.

5. Area is the number of square units it takes to _____ a polygon.

6. The area of a triangle is _____ times _____ times _____.

Use the bank of answers to find the areas for the figures below.

510yd^2	450yd^2	48cm^2	96cm^2
18in^2	81in^2	80in^2	160in^2

7.

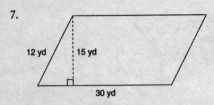

8.

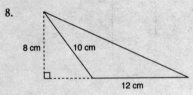

9.

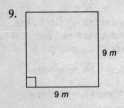

10.

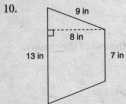

Circles

The **radius**, r, of a circle is the segment whose endpoints are the center of the circle and any point on the circle. The **diameter**, d, of a circle is a segment that passes through the center of the circle and whose endpoints are both on the circle. The diameter's length is twice the length of the radius.

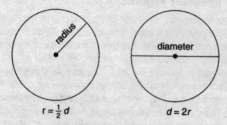

$r = \frac{1}{2}d$ $d = 2r$

REMEMBER THIS!

The perimeter of a circle is called the **circumference**. The circumference of a circle is calculated with the formula $C = 2\pi r$, or, alternately, $C = \pi d$.

The area of a circle is calculated with the formula $A = \pi r^2$.

For example, to find the circumference of a circle with diameter of 34 inches, use the formula $C = \pi d = 3.14 \times 34 = 106.76$ inches. To find the area, first calculate the radius, which is one-half of the diameter, or 17 inches. Then use the formula $A = \pi r^2$, or $A = 3.14 \times 17 \times 17 = 907.46$ in^2.

Practice 3

Answer true or false to the following statements. Answers and explanations are located at the end of the chapter.

11. **T   F**   The length of the diameter of a circle with radius 32 cm is 16 cm.

12. **T F** Using $\pi = 3.14$, the circumference of a circle with radius 10 inches is 62.8 inches.

13. **T F** The area of a circle with radius 8 cm is twice the area of a circle with radius of 4 cm.

Area of Irregular Figures

Perhaps you have used the concept of area with home improvement projects. If you have ever wanted to wallpaper the walls of a room, for example, you need to calculate the area of the walls in the room. Seldom are rooms shaped as perfect rectangles, and so you sometimes need to calculate the area of an irregular shaped polygon. To accomplish this task, break the shapes up into recognizable polygons, and then calculate the area of each section; add these areas together to find the total area. For example, look at the irregular figures below and identify the recognizable polygon shapes:

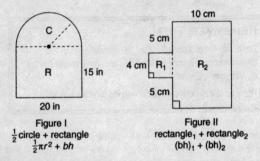

Figure I
$\frac{1}{2}$ circle + rectangle
$\frac{1}{2}\pi r^2 + bh$

Figure II
rectangle$_1$ + rectangle$_2$
$(bh)_1 + (bh)_2$

To find the area of Figure I above, calculate $\frac{1}{2}$ of the area of a circle, and add to this the area of a rectangle. So the area of Figure I is

$(\frac{1}{2} \times \pi \times 10 \times 10) + (20 \times 15) = 157 + 300 = 457$ cm^2. To find the

area of Figure II, add together the area of the two rectangles. The smaller rectangle has an area of $4 \times 5 = 20$ and the larger rectangle has an area of $10 \times (5 + 4 + 5) = 10 \times 14 = 140$. In total, the area is 160 square units.

Area of Shaded Regions

At times, you are presented with the problem of finding the area of shaded regions, as shown below:

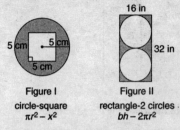

Figure I	Figure II
circle-square	rectangle-2 circles
$\pi r^2 - x^2$	$bh - 2\pi r^2$

In these examples, the shaded region is part of the outer portion of a recognizable figure; the un-shaded area is a second recognizable figure in the interior. One way to calculate this area is to find the area of the entire outer shape, and then subtract the area of the inner un-shaded shape. For example, in Figure I above, the outer figure is a circle of radius 5 cm and the inner polygon is a square with side length of 5 cm. The area of the shaded region is (the area of the circle) − (the area of the square). By substituting in the known formulas, and using the value of 3.14 for π, the area is $(\pi r^2) - (s^2) = (3.14 \times 5 \times 5) - (5 \times 5)$. This simplifies to $78.5 - 25 = 53.5$ cm^2.

In Figure II above, the outer figure is a rectangle with dimensions of 16 inches and 32 inches, and the inner figures are two congruent circles of radius 8 inches. Since the circles touch both sides of the rectangle, their diameters are 16 inches; therefore the radius is 8 inches. The area of the shaded region is (area of the rectangle) − (2 times the area of the circle). Again, using 3.14 for π, the area is $(16 \times 32) - (2 \times \pi \times 8 \times 8) = 512 - 401.92 = 110.08$ inches^2.

Practice 4

14. Find the area of the following irregular figure:

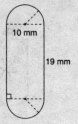

15. Find the area of the following irregular figure:

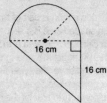

16. Find the area of the shaded region:

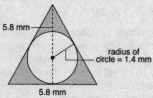

17. Find the area of the shaded region:

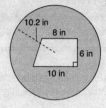

Surface Area

The measure of surface area applies to three-dimensional solids.

> **MATH SPEAK**
>
> A **prism** is a three-dimensional solid with two congruent bases, and any number of other faces that are all rectangles.
>
> A **cylinder** has two congruent circular bases, with one rectangular face wrapped around the bases.

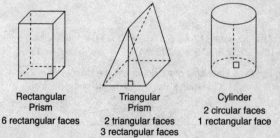

Rectangular Prism	Triangular Prism	Cylinder
6 rectangular faces	2 triangular faces	2 circular faces
	3 rectangular faces	1 rectangular face

These solids have faces that are made up of common polygons.

> ## **REMEMBER THIS!**
> The surface area of any three-dimensional solid is the sum of the areas of its faces.

To find the surface area of any solid, calculate the area of each of the faces, and then add them all together. The units for surface area are square units.

The most common solid is the rectangular solid, which has six rectangular faces. For this solid, all of the opposite faces are congruent, so the surface area can be calculated as (2 × front face) + (2 × left face) + (2 × top face). The formula for each of these faces is bh. To find the surface area of the rectangular solid below, substitute in the given dimensions:

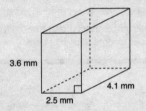

The surface area is $(2 \times 2.5 \times 3.6) + (2 \times 3.6 \times 4.1) + (2 \times 2.5 \times 4.1)$ $= 18 + 29.52 + 20.5 = 68.02$ mm².

To find the surface area of a cylinder, use the formula $SA = 2\pi r^2 + 2\pi rh$, where r is the radius and h is the height of the cylinder.

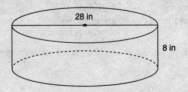

First, recognize that the diameter is given in the figure. Calculate the radius by taking one-half of the diameter, or 14 in. Then substitute in this radius and the height of 8 in. The surface area is $(2 \times \pi \times 14 \times 14) + (2 \times \pi \times 14 \times 8) = 1,230.88 + 703.36 = 1,934.24$ in^2.

Practice 5

Find the surface area of the solids below. Answers and explanations are located at the end of the chapter.

18.

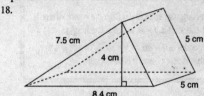

19.

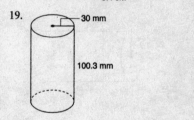

20.

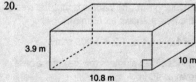

Volume

The **volume** of a three-dimensional solid is the amount of cubic units needed to fill the solid. The volume of a prism or a cylinder is the area of one of the bases, B, multiplied by the height, h.

For a prism, the formula for volume is the area of the base, B, multiplied by the height. So for a triangular prism, the volume is the area of the triangular base $\left(\frac{1}{2}bh\right)$, multiplied by the height of the prism. The two most common solids to find the volume of are the rectangular prism and the cylinder.

REMEMBER THIS!

The volume of a rectangular solid is $V = Bh$, or $V = lwh$, where B is the area of the base, that is, $l \times w$, and h is the height.

The volume of a cylinder is $V = Bh$, or $V = \pi r^2 h$, where B is the area of the base, that is $\pi \times r^2$, and h is the height.

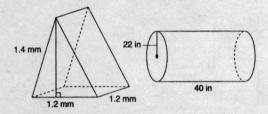

The left-hand figure is a triangular prism. The volume is thus the area of the base, a triangle, times the height, shown as 1.2 mm.

$$V = \left(\frac{1}{2} \times b \times h\right) \times h = \left(\frac{1}{2} \times 1.2 \times 1.4\right) \times 1.2 = 1.008 \text{ mm}^3.$$

The right-hand figure is a cylinder. The volume is the area of the base, a circle, times the height, shown as 40 in.

$$V = \pi \times 22^2 \times 40 = 60{,}790.4 \text{ in}^3.$$

Volume of a Pyramid and a Cone

The pyramid and the cone are special three-dimensional solids that are related to the prism and the cylinder. The pyramid is a rectangular solid in which instead of two congruent faces, there is a point opposite to the base of the solid. The cone is similarly related to the cylinder. Some examples are shown on the next page.

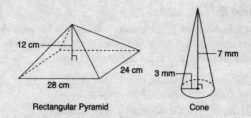

12 cm

24 cm

28 cm

Rectangular Pyramid

7 mm

3 mm

Cone

> ## REMEMBER THIS!
> The volume of a pyramid or a cylinder is $V = \frac{1}{3}Bh$.
>
> The volume of a pyramid is $V = \frac{1}{3}lwh$.
>
> The volume of a cone is $V = \frac{1}{3}\pi r^2 h$.

To find the volume of the rectangular pyramid shown above, multiply $\frac{1}{3}Bh$, where B is the area of the rectangular base (which is 28×24, or 672 cm^2) and h is the height, or 12 cm. The volume is therefore $\frac{1}{3} \times 672 \times 12 = 2{,}688$ cm^3. The volume of the cone shown above is again $\frac{1}{3}Bh$. In this case, B is the area of the circular base, or πr^2, which is $3.14 \times 3 \times 3 = 28.26$ mm^2, and h is the height shown of 7 mm. The volume is thus $\frac{1}{3} \times 28.26 \times 7 = 65.94$ mm^3.

Practice 6

Use the number bank of cubic units to find the volumes of the
three-dimensional solids, to the nearest tenth. Answers and
explanations are located at the end of the chapter.

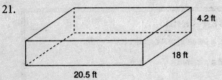

42.7ft³	1549.8ft³	387.7cm³
3799.7cm³	949.9cm³	1040.4mm³
3121.2mm³	35196.6mm³	11732.2mm³

21.

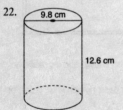

4.2 ft

18 ft

20.5 ft

22.

9.8 cm

12.6 cm

23.

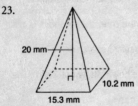

20 mm

10.2 mm

15.3 mm

24.

32.4"

18.6"

THE PYTHAGOREAN THEOREM

The right triangle is a very common polygon found in everyday life. One of the most widely used theorems in mathematics is the Pythagorean theorem, which is based upon right triangles.

MATH SPEAK

The **hypotenuse** of a right triangle is the side that is opposite from the right angle.

The **legs** of a right triangle are the sides that make up the right angle.

The Pythagorean theorem states that in every right triangle, the sum of the squares of the legs is equal to the square of the hypotenuse. This is commonly shown as $a^2 + b^2 = c^2$, where a and b are the lengths of the legs of the right triangle and c is the length of the hypotenuse. The converse of the Pythagorean theorem is also true:

If a triangle has sides whose lengths follow the relationship that $a^2 + b^2 = c^2$, then the triangle is a right triangle.

The Pythagorean theorem is used to find the missing length of a side of a triangle when any two of the lengths are known. For example, given that a right triangle has legs with length of 12 mm and 5 mm, the length of the hypotenuse, c, is found with the theorem $a^2 + b^2 = c^2$. Substitute in 12 and 5 for a and b to get $12^2 + 5^2 = c^2$. Simplify: 144 + 25 = c^2, or 169 = c^2. The length of the hypotenuse is the square root of 169, or 13 mm.

FLASHBACK

Square roots were covered in chapter 7, Powers and Roots.

The converse of the theorem is a convenient way to test whether a triangle is a right triangle. For example, if you are given that a triangle has sides of 5 in., 8 in., and 13 in., you can be assured that the triangle is NOT a right triangle, because $5^2 + 8^2 \neq 13^2$, that is $25 + 64 = 89$, not 169. However, a triangle with sides of 3 cm, 4 cm, and 5 cm IS a right triangle, because $3^2 + 4^2 = 5^2$, that is $9 + 16 = 25$.

Practice 7

Try these problems based on the Pythagorean theorem. To master mathematics, practice is the only route to success! Answer true or false to the following statements.

25. **T F**   A triangle with sides of 7, 10 and 17 is a right triangle.

26. **T F**   A triangle with sides of 9, 12 and 15 is a right triangle.

27. **T F**   If the two legs of a right triangle are 10 and 24, then the hypotenuse is 26.

28. **T F**   If one leg of a right triangle is 3 and the hypotenuse of the right triangle is 5, then the other leg is $\sqrt{34}$, or 5.83.

TRIGONOMETRY

Another useful application of right triangle measurement is trigonometry. Trigonometry deals with the acute angles of a right triangle, and the ratio of the lengths of the sides.

> **FLASHBACK**
>
> The concept of ratio is explored in chapter 4, Ratio and Proportion.

To explore the uses for trigonometry, you must first be able to identify, in a right triangle, the side that is adjacent to an acute angle and the side that is opposite to an acute angle.

> ## MATH SPEAK
>
> The side of a right triangle that is **adjacent** to an acute angle in the triangle is the side that forms this acute angle that is not the hypotenuse.
>
> The side of a right triangle that is **opposite** to an acute angle in the triangle is the side that does not form the acute angle.

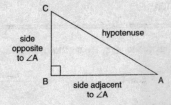

As stated above, the trigonometric ratios compare the lengths of the sides of the right triangle in relationship to one of the acute angles. These ratios are formed as fractions, and are then sometimes converted to decimal equivalents.

> ## REMEMBER THIS!
>
> The **sine (sin)** of an acute angle is
>
> $\dfrac{\text{length of the opposite side}}{\text{length of the hypotenuse}}$, or simply $\dfrac{O}{H}$.
>
> The **cosine (cos)** of an acute angle is
>
> $\dfrac{\text{length of the adjacent side}}{\text{length of the hypotenuse}}$, or simply $\dfrac{A}{H}$.
>
> The **tangent (tan)** of an acute angle is
>
> $\dfrac{\text{length of the opposite side}}{\text{length of the adjacent side}}$, or simply $\dfrac{O}{A}$.
>
> Remember this as the popular mnemonic $S\dfrac{O}{H} \cdot C\dfrac{A}{H} \cdot T\dfrac{O}{A}$ or SOH CAH TOA.

Study the figure below and verify the trigonometric ratios shown:

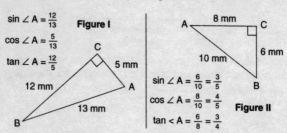

$\sin \angle A = \frac{12}{13}$ **Figure I**

$\cos \angle A = \frac{5}{13}$

$\tan \angle A = \frac{12}{5}$

12 mm

C 5 mm

A

13 mm

B

A 8 mm C

10 mm 6 mm

B

$\sin \angle A = \frac{6}{10} = \frac{3}{5}$

$\cos \angle A = \frac{8}{10} = \frac{4}{5}$

$\tan < A = \frac{6}{8} = \frac{3}{4}$ **Figure II**

It is important to realize that each acute angle in the right triangle has its own unique values for sine, cosine, and tangent. For example, in right triangle I above, the $\sin \angle B = \dfrac{5}{13}$ and the $\tan \angle B = \dfrac{5}{12}$.

Right triangles with corresponding acute angles that are congruent are similar, and therefore the sides are in proportion.

FLASHBACK

Similarity and proportions were covered in chapter 4, Ratio and Proportion.

Because of similarity, the sine, cosine, and tangent of an acute angle is always in the same ratio. If you know the degree measure of an acute angle, any one of the three trigonometric ratios is given on a scientific calculator with the press of a button, (sin,cos,tan). For example, if you enter cos(60°) on a calculator, the result will be 0.5. This decimal is equivalent to $\frac{1}{2}$, so the ratio of the length of the adjacent side to the length of the hypotenuse (of the 60° angle) is 1 to 2. Likewise, if you are trying to find the angle measure of an angle when you know the side lengths, you can also find this information on a calculator. To do this, you convert the fraction to its decimal equivalent and then use the appropriate inverse trigonometric key, (sin⁻¹,cos⁻¹,tan⁻¹). For example, to find the measure of an angle whose sine is $\frac{1}{2} = 0.5$, enter into a calculator sin⁻¹ 0.5, and the calculator will give you the measure as 30°. Note that some calculators may require you to enter this as 0.5 sin⁻¹. Test your specific calculator to be sure.

Practice 8

Remember, practice is the key to mastery! Match each problem in the left-hand column with the correct value in the right-hand column. Some right-hand column values will be used more than once, and some may not be used at all. Base your answers on the following diagram.

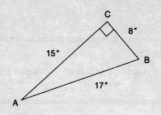

29. cos ∠A

30. sin ∠B

31. tan ∠B

32. tan ∠A

33. cos ∠B

1. $\dfrac{8}{15}$

2. $\dfrac{8}{17}$

3. $\dfrac{15}{17}$

4. $\dfrac{15}{8}$

Applications of the Pythagorean Theorem and Trigonometry

If you are given a problem situation that involves a right triangle, you will almost certainly use either the Pythagorean theorem or trigonometry in order to solve the problem. Sometimes it seems difficult to know which concept to apply. It is actually quite simple to make this decision. If the problem involves one of the angles of the triangle, then use trigonometry; if the problem situation involves the side lengths only, then use the Pythagorean theorem.

The following are examples that use these right triangle concepts. Note that the solution to each of these real world examples involves identifying the right triangle and then deciding which concept to apply.

What is the length of the diagonal of the following rectangle?

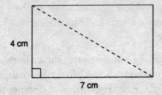

The diagonal is the hypotenuse of the right triangle with legs of length 4 cm and 7 cm. Use the Pythagorean theorem to solve for the hypotenuse: $a^2 + b^2 = c^2$. $4^2 + 7^2 = c^2$; $16 + 49 = c^2$; $65 = c^2$, so $c = \sqrt{65} \approx 8.1$ cm.

As shown in the following diagram, a boat sights the top of a lighthouse when it is 100 feet from the base of the lighthouse. The angle of elevation is 45°. How tall is the lighthouse?

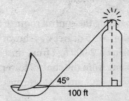

The height of the lighthouse is the side that is opposite to the 45° angle, and the horizontal distance (100 feet) is the side that is adjacent to the angle. Use the tangent ratio to solve for the height of the lighthouse, represented as x. The $\tan 45° = \dfrac{\text{opposite}}{\text{adjacent}}$; $\tan 45° = \dfrac{x}{100}$; $\tan 45° \times 100 = x$. Since the tangent of 45° = 1, the height of the lighthouse is 100 feet.

Practice Set 9

Try the following problem situations that involve right triangles and measurements. Answers and explanations are located at the end of the chapter.

34. A man is climbing a mountain, as shown in the diagram. The slope of the mountain is 35°. What is his change in elevation, represented as x, when his horizontal distance from the starting point is 1,000 feet? Give your answer to the nearest tenth.

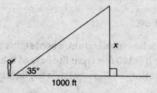

35. A plane is approaching the control room at an airport at an angle of depression of 15 degrees, as shown below. If the altitude of the plane is 2,000 feet from the control room, how far away is the plane from the control tower, as represented by x in the figure? Give your answer to the nearest tenth.

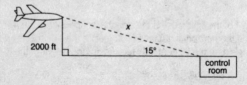

36. What is the length of the diagonal of a rectangle with sides of 12.6 cm and 8.4 cm?

37. A wire is needed to support a telephone pole, as shown in the drawing below. The wire is to be attached at a height of 20 feet on the pole, and the angle is to be 70 degrees. How long is the required length of wire, to the nearest hundredth?

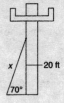

SUMMARY

In conclusion, here are seven things about measurement you should take from this chapter:

- Perimeter is the sum of the distance around the exterior of a polygon. Circumference is the perimeter of a circle.

- Area is the amount of square units needed to cover a geometric figure. Know the various formulas for common geometric figures.

- Surface area is the sum of the areas of the faces of a three dimensional geometric solid.

- Volume is the amount of cubic units needed to fill a geometric solid. For prisms and cylinders, it is found by multiplying the area of the base times the height.

- The volume of a pyramid is one-third the volume of a rectangular solid. The volume of a cone is one-third the volume of a cylinder.

- For right triangles, the sum of the squares of the legs is equal to the square of the hypotenuse. This is the Pythagorean theorem, that is $a^2 + b^2 = c^2$.

- For right triangles, many measurements can be found by trigonometry, which can be recalled by the mnemonic $S\dfrac{O}{H} - C\dfrac{A}{H} - T\dfrac{O}{A}$.

Practice Answers and Explanations

1.   253 mm

Add up the lengths of the sides of the triangle. The height of the triangle is shown in the figure, but is not used to calculate perimeter.

2.   64 m

The perimeter of a square is four times the length of one side. $16 \times 4 = 64$ m.

3.   24 in

The perimeter of this parallelogram is $(2 \times 7) + (2 \times 5) = 14 + 10 = 24$ in.

4.   square

5.   cover

6.   one-half × base × height

7.   450 yd²

Area of a parallelogram is base times height. $A = 30 \times 15 = 450$.

8.   48 cm²

Area of a triangle is is $\frac{1}{2}bh$. So $A = \frac{1}{2} \times 12 \times 8$

$A = 6 \times 8$

$A = 48$

9.   81 m²

$A = l \times w$; Area $= 9m \times 9m$; Area $= 81 m^2$

10. 80 in²

The area of a trapezoid is $\frac{1}{2}(b_1 + b_2)h$. $A = \frac{1}{2} \times (7 + 13) \times 8 = 80$.

11. False

$D = 2r$, so the diameter would be 2×32, or 64.

12. True

$C = 2\pi r$; $C = 2 \times 3.14 \times 10$; $C = 62.8$ in.

13.   False

Since the radius is squared in the area formula, the larger circle will be four times the area of the smaller circle.

14.   268.5 mm^2

The figure contains two congruent halves of a circle, so it is the area of a circle, plus the area of a rectangle. $A = \pi r^2 + bh$. The radius is one-half of the diameter. The diameter is the same as a side of the rectangle.

$A = (3.14)(5^2) + (10)(19)$

$A = 78.5 + 190$

$A = 268.5$ mm^2

15.   228.48 cm^2

The figure is one-half of a circle, plus a triangle.

$A = 0.5 \times \pi \times r^2 + \dfrac{1}{2} \times b \times h.$

$A = 5 \times 3.14 \times (8)^2 + \dfrac{1}{2}(16)(16)$

$A = 100.48 \times 128$

$A = 228.48$ cm^2

16.   10.6656 mm^2

The figure is a triangle, minus the unshaded circle.

$A = \dfrac{1}{2} \times b \times h - \pi \times r^2.$

$A = \dfrac{1}{2}(5.8)(5.8) - 3.14(1.4)2$

$A = 16.82 - 6.1544$

$A = 10.6656$ mm^2

17.   272.6856 in^2

The figure is a circle, minus the unshaded trapezoid.

$A = \pi \times r^2 - \dfrac{1}{2} \times (b_1 + b_2) \times h.$

$A = 3.14(10.2)^2 \times \dfrac{1}{2} \times (8 + 10)(6)$

$A = 326.69 - 54$

$A = 272.6856 \text{ in}^2$

18. 138.1 cm²

This is a triangular prism, with two congruent triangular faces and three rectangular faces. Find each of these areas and add them together.

$SA = 2 \times \dfrac{1}{2} \times 4 \times 8.4 + 5 \times 5 + 7.5 \times 5 + 5 \times 8.4$

$SA = 33.6 + 25 + 37.5 + 42$

$SA = 138.1 \text{ cm}^2$

19. 24,548.52 mm²

This is a cylinder, so the surface area is $(2 \times \pi \times r^2) + (2 \times \pi \times r \times h)$.

$SA = (2 \times 3.14 \times 30^2) + (2 \times 3.14 \times 30 \times 100.3)$

$SA = 5,652 + 18,896.52$

$SA = 24,548.52 \text{ mm}^2$

20. 378.24 m²

This is a rectangular prism. There are three pairs of congruent rectangular faces.

$SA = (2 \times 3.9 \times 10.8) + (2 \times 10 \times 10.8) + (2 \times 3.9 \times 10)$

$SA = 84.24 + 216 + 78$

$SA = 378.24 \text{ m}^2$

21. 1,549.8 ft³

The volume of a rectangular solid is $V = l \times w \times h$.

$V = 20.5 \times 1.8 \times 4.2$

$V = 1,549.8 \text{ ft}^3$

22. 949.9 cm³

The volume of a cylinder is $V = \pi \times r^2 \times h$.

$V = 3.14 \times (4.9)^2 \times 12.6$

$V = 949.9 \text{ cm}^3$

23. 1,040.4 mm^3

The volume of a rectangular pyramid is $V = \frac{1}{3} \times l \times w \times h$.

$V = \frac{1}{3} \times 10.2 \times 15.3 \times 20$

$V = 1,040.4$ mm^3

24. 11,732.2 mm^3

The volume of a cone is $V = \frac{1}{3} \times \pi \times r^2 \times h$.

$V = \frac{1}{3} \times 3.14 \times (18.6)^2 \times 32.4$

$V = 11,732.2$ mm^3

25. False

$7^2 + 10^2 \neq 17^2$

26. True

$9^2 + 12^2 = 15^2$

27. True

$10^2 + 24^2 = 26^2$

28. False

$3^2 + b^2 = 5^2$

$9 + b^2 = 25$

$b^2 = 16$

$b = 4$

$4 \neq 5.83$

29. 3

The cosine of angle $\angle A$ is $\dfrac{\text{adjacent side}}{\text{hypotenuse}} = \dfrac{15}{17}$.

30. 3

The sine of angle $\angle B$ is $\dfrac{\text{opposite side}}{\text{hypotenuse}} = \dfrac{15}{17}$.

31. 4

The tangent of angle $\angle B$ is $\dfrac{\text{opposite side}}{\text{adjacent side}} = \dfrac{15}{8}$.

32. 1

The tangent of angle $\angle A$ is $\dfrac{\text{opposite side}}{\text{adjacent side}} = \dfrac{8}{15}$.

33. 2

The cosine of angle $\angle B$ is $\dfrac{\text{adjacent side}}{\text{hypotenuse}} = \dfrac{8}{17}$.

34. 700.2 feet

The rise in elevation is the side that is opposite to the angle of 35°; the horizontal distance, 1,000 feet, is the side that is adjacent to the angle. Set up the ratio for the tangent of 35°, by representing the altitude as x: $\tan 35° = \dfrac{x}{1,000}$. The altitude, x, is thus $1,000 \times \tan 35° = 700.2$ feet, rounded to the nearest tenth.

35. 7,727.4 feet

The distance in the figure, x, is the hypotenuse of the right triangle. The altitude, 2,000 feet, is the side opposite to the angle of 15°. Use the sine ratio to solve for x: $\sin 15° = \dfrac{2,000}{x}$. To find the value of x, divide 2,000 by the $\sin 15°$, $\dfrac{2,000}{\sin 15°} = 7,727.4$ feet.

36. $\sqrt{229.32} \approx 15.14$ cm

The diagonal is the hypotenuse of a right triangle with legs of length 12.6 cm and 8.4 cm. Use the Pythagorean theorem to solve for the hypotenuse.

37. 21.28 feet

The length of the wire is represented as x in the diagram. This is the hypotenuse of a right triangle. The height on the pole (20 feet) is the side that is opposite to the 70° angle. Use the sine ratio: $\sin 70° = \dfrac{20}{x}$, so therefore $x = \dfrac{20}{\sin 70°}$, which is 21.28 feet, to the nearest hundredth.

CHAPTER 9 TEST

Now that you have read over and studied the concepts of measurement, try the following questions to test your skills and knowledge. The answers and explanations that follow will help clarify new concepts.

1. What is the perimeter of a square with sides of 57 mm?
   (A) 114 mm   (B) 324.9 mm   (C) 228 mm
   (D) 61 mm   (E) 59 mm

2. Every time its wheel turns, a bicycle travels a distance equal to the circumference of the wheel. If a bicycle wheel's diameter is 0.5 m, approximately how many turns will the wheel have to make for the bike to travel 100 meters?
   (A) 200 turns   (B) 20 turns   (C) 64 turns
   (D) 50 turns   (E) 500 turns

3. Gracie wants to build a fence around his swimming pool area. The area is shaped like a rectangle, with dimensions of 20 ft by 14 ft. How much fencing is needed?
   (A) 280 ft   (B) 68 ft   (C) 80 ft
   (D) 34 ft   (E) 596 ft

4. What is the area of a circle with diameter of 58 mm?
   (A) 91.06 mm^2 (B) 13,456 mm^2 (C) 841 mm^2
   (D) 364.24 mm^2 (E) 2,640.74 mm^2

5. Find the area of the following polygon:

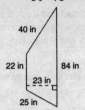

(A) 194 in²   (B) 2,438 in²   (C) 171 in²
(D) 1,219 in²   (E) 1,932 in²

6. Find the area of the triangle:

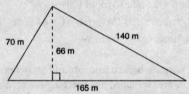

(A) 11,550 in²   (B) 10,890 in²   (C) 5,445 in²
(D) 5,775 in²   (E) 375 in²

7. Find the surface area:

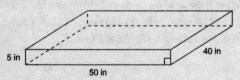

(A) 10,000 in²   (B) 2,450 in²   (C) 95 in²
(D) 2,000 in²   (E) 4,900 in²

8. Find the surface area:

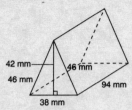

(A) 150,024 mm²   (B) 75,012 mm²   (C) 13,018 mm²

(D) 13,816 mm²   (E) 15,412 mm²

9. How much paint is needed to completely cover a box with dimensions of 2 feet by 1.5 feet by 3 feet?

(A) 27 ft²   (B) 9 ft²   (C) 6.5 ft²

(D) 13.5 ft²   (E) 13 ft²

10. The school district wants to install sod in the interior of the track field. How much sod is needed if the dimensions and shape are as shown below:

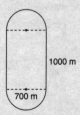

(A) 4,198 m²   (B) 702,198 m²   (C) 1,469,300 m²

(D) 708,792 m²   (E) 1,084,650 m²

11. Find the area:

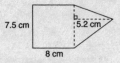

(A) 79.5 cm²   (B) 99 cm²   (C) 101.6 cm²

(D) 80.8 cm²   (E) 36.2 cm²

12. Find the area:

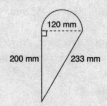

(A) 19,632 mm²   (B) 12,188.4 mm²   (C) 17,652 mm²

(D) 29,652 mm²   (E) 18,120 mm²

13. Find the area of the shaded region:

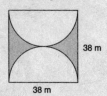

(A) 1,324.68 m²   (B) 310.46 m²   (C) 1,205.36 m²

(D) 1,444 m²   (E) 2,577.54 m²

14. Find the area of the shaded region:

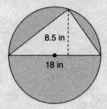

(A) 198.9 in²   (B) 330.84 in²   (C) 101.34 in²

(D) 177.84 in²   (E) 220.34 in²

15. What is the volume of the following cylinder?

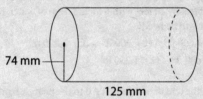

(A) 2,149,330 mm³   (B) 58,090 mm³

(C) 35,694.64 mm³   (D) 4,298,660 mm³

(E) 684,500 mm³

16. What is the volume of the triangular prism?

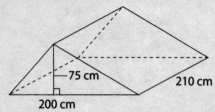

(A) 49,500 cm³ (B) 1,575,000 cm³ (C) 3,150,000 cm³

(D) 72,750 cm³ (E) 81,000 cm³

17. Dave wants to put an ice rink in his backyard. The rink will be 30 feet by 20 feet with 6 inches deep of water. How much water will he need?

(A) 3,600 ft³ (B) 300 ft³ (C) 7,200 ft³

(D) 600 ft³ (E) 1,250 ft³

18. Find the length of segment AB to the nearest tenth.

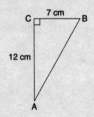

(A) 6.2 cm (B) 9.5 cm (C) 13.9 cm

(D) 96.5 cm (E) 19 cm

19. Find the length of segment CB to the nearest tenth.

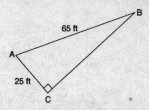

(A) 69.6 ft   (B) 40 ft   (C) 8.9 ft   (D) 60 ft   (E) 42 ft

20. A car travels directly west for 50 miles, and then travels directly north for 15 miles. How much shorter, to the nearest tenth, would the trip have been if the car had drove diagonally NW, the shortest route?

(A) 12.8 miles   (B) 52.2 miles   (C) 65 miles

(D) 11.4 miles   (E) 47.4 miles

21. How far is it across the pond (*x* in the diagram)?

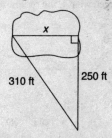

(A) 110 ft   (B) 60 ft   (C) 16,800 ft

(D) 120 ft   (E) 183.3 ft

Use the following diagram for questions 22 and 23.

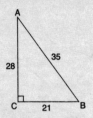

22. What is the sin ∠A?

    (A) $\dfrac{28}{35}$ (B) $\dfrac{35}{21}$ (C) $\dfrac{21}{35}$ (D) $\dfrac{21}{28}$ (E) $\dfrac{28}{21}$

23. What is the degree measure of ∠B?

    (A) 36.9° (B) 0.01° (C) 48.6° (D) 53.1° (E) 41.4°

24. An archer shoots an arrow at a 6° angle of elevation. If she is standing 30 feet from the target and the arrow lands in the center of the target, what is the radius of the target, to the nearest tenth?

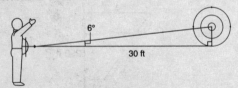

    (A) 5 ft (B) 3.2 ft (C) 180 ft (D) 50.5 ft (E) 30 ft

Answers and Explanations

1. C

Because all sides of a square are equal, the perimeter of a square is 4 times the length of one side. 4 × 57 = 228 mm.

2.   C

The circumference of the wheel is how far the bicycle will travel with each turn of the wheel. The circumference of a circle is $\pi \times d$; the distance traveled with one turn is $3.14 \times 0.5 = 1.57$ m. Divide this into the total distance of 100 and $100 \div 1.57 = 63.7$, or 64 turns.

3.   B

The perimeter of a rectangle is 2 times the length plus 2 times the width. $P = (2 \times 20) + (2 \times 14) = 40 + 28 = 68$ ft.

4.   E

The area of a circle is found using the formula $\pi \times r^2$. The radius is one-half of the diameter, or 29 mm. The area is thus $3.14 \times 29 \times 29 = 2,640.74$ mm^2.

5.   D

The formula for the area of a trapezoid is $A = \frac{1}{2} \times (b_1 + b_2) \times h$. The bases are the lengths of the parallel sides, or 22 in. and 84 in.. The height is 23 in. Therefore, $A = 0.5 \times (22 + 84) \times 23 = 1,219$ in^2.

6.   C

The area of a triangle is found by the formula $A = \frac{1}{2} \times b \times h$.

For this triangle, the base is 165 m and the height is 66 m. The area is $0.5 \times 165 \times 66 = 5,445$ m^2.

7.   E

The surface area is the sum of the area of each face. There are three pairs of congruent rectangular faces on a rectangular solid, and the area of each face is base times height. The surface area is $2 \times (50 \times 5) + 2 \times (50 \times 40) + 2 \times (40 \times 5) = 4,900$ in^2.

8.   D

A triangular prism has five faces: 2 congruent triangular bases, and three rectangular faces. The surface area is $2 \times (\frac{1}{2} \times 38 \times 42) + (94 \times 46) + (94 \times 46) + (94 \times 38) = 13,816$ mm^2.

9. A

This is a surface area problem. There are three pairs of congruent rectangular faces on a rectangular solid. The surface area is thus $2 \times (2 \times 1.5) + 2 \times (2 \times 3) + 2 \times (1.5 \times 3) = 27$ ft^2.

10. E

The track field is an irregular shaped polygon featuring 2 congruent halves of a circle with a radius of 350m, (or one complete circle), and one rectangle with dimensions of 700 m by 1000 m. Find the area of the rectangle and the circle and add them together. Area $= \pi r^2 + bh$. $A = (3.14 \times 350 \times 350) + (700 \times 1000) = 1,084,650$ m^2.

11. A

The polygon is a rectangle and a triangle. Find the area of each of shape and add them together. $A = \frac{1}{2}bh + bh$. $A = \frac{1}{2} \times (7.5 \times 5.2) + (7.5 \times 8) = 79.5$ cm^2.

12. C

The irregular shaped polygon is one-half of a circle plus a triangle. $A = \frac{1}{2}\pi r^2 + \frac{1}{2}bh$. The radius of the circle is one-half of the diameter shown (the width of the rectangle), or 60 mm.

$A = \left(\frac{1}{2} \times 3.14 \times 60 \times 60\right) + \frac{1}{2}(120 \times 200) = 17,652$ mm^2.

13. B

The area of the shaded region is the area of the outer figure (a square), minus the area of the unshaded inner region (two congruent halves of a circle, or one complete circle). The area of the shaded is thus $bh - \pi r^2$. $A = (38 \times 38) - (3.14 \times 19 \times 19) = 310.46$ m^2.

14. D

The area of the shaded region is the area of the outer figure, a circle, minus the area of the unshaded inner region, a triangle. The area is thus $\pi r^2 - \frac{1}{2}bh$; $A = (3.14 \times 9 \times 9) - (\frac{1}{2} \times 18 \times 8.5) = 177.84$ in^2.

15. A

The volume of a cylinder is $\pi r^2 h$. The volume is thus $3.14 \times 74 \times 74 \times 125 = 2{,}149{,}330$ mm^3.

16. B

The volume of a triangular solid is the area of the base (a triangle) times the height. $V = (\frac{1}{2}bh) \times h$. $V = (\frac{1}{2} \times 200 \times 75) \times 210 = 1{,}575{,}000$ cm^3.

17. B

This is a rectangular solid, whose height is 0.5 feet (6 inches is one-half of a foot). $V = 30 \times 20 \times 0.5 = 300$ ft^3.

18. C

Segment AB is the hypotenuse of the right triangle $\triangle$ACB. Use the Pythagorean theorem, with the hypotenuse equal to c: $a^2 + b^2 = c^2$. $7^2 + 12^2 = c^2$; $49 + 144 = c^2$; $193 = c^2$; $c = \sqrt{193} = 13.9$ cm.

19. D

Segment CB is a leg of a right triangle with hypotenuse of 65. Use the Pythagorean theorem to solve for the leg, represented as b: $a^2 + b^2 = c^2$. $25^2 + b^2 = 65^2$; $625 + b^2 = 4{,}225$; $b^2 = 4{,}225 - 625$; $b^2 = 3{,}600$; $b = \sqrt{3600} = 60$ ft.

20. A

The route of the car forms a right triangle, since the car drove directly west and then north. The shortest distance is the hypotenuse of the right triangle. Use the Pythagorean theorem, with the hypotenuse equal to c: $a^2 + b^2 = c^2$; $50^2 + 15^2 = c^2$; $2500 + 225 = c^2$; $2{,}725 = c^2$. $c = \sqrt{2725}$, or approximately 50.2 miles. The original trip was $50 + 15 = 65$ miles. Since this route would be 50.2 miles, it would be $65 - 50.2 = 12.8$ miles shorter.

21. E

The distance across the pond, represented by x, is a leg of the right triangle with hypotenuse 310 feet. Use the Pythagorean theorem, $a^2 + b^2 = c^2$; $250^2 + x^2 = 310^2$; $62{,}500 + x^2 = 96{,}100$; $x^2 = 96{,}100 - 62{,}500$; $x^2 = 33{,}600$. $x = \sqrt{33600}$, or approximately 183.3 feet.

22. C

The sine of angle $\angle A$ is the ratio of $\dfrac{\text{length of the opposite side}}{\text{length of the hypotenuse}} = \dfrac{21}{35}$.

23. D

Use trigonometry to find the degree measure of the angle. The sine of angle $\angle B$ is the ratio of $\dfrac{\text{length of the opposite side}}{\text{length of the hypotenuse}} = \dfrac{28}{35}$. The angle measure is therefore $\sin^{-1}\left(\dfrac{28}{35}\right)$, or angle $\angle B$ is approximately $53.1°$.

24. B

The diagram is a right triangle situation. An angle is known, and only one side of the triangle, (the distance from the archer to the target), is adjacent to the angle of $6°$. You need to use trigonometry to solve the problem. The radius of the target is the side that is opposite to the angle of $6°$. Use the tangent ratio to solve for x, the opposite side.

$\text{Tan } 6° = \dfrac{\text{length of the opposite side}}{\text{length of the adjacent side}} = \dfrac{x}{30}$. So $x = \tan 6° \times 30$, or x is approximately 3.2 feet.

CHAPTER 10

Coordinate Geometry

Begin your journey through coordinate geometry and its applications with this 10-question Building Block Quiz. This quiz includes three review questions from the previous chapter on measurement, and a pre-assessment on the new materials presented in this chapter. Complete answer explanations following the quiz will assist you in your study.

BUILDING BLOCK QUIZ

1. What is the area in square inches of a triangle with a base of 6 inches and a height of 8 inches?

   (A) 12   (B) 16   (C) 24   (D) 36   (E) 48

2. The length of the hypotenuse of a right triangle is 78 cm. One of the legs has a measure of 72 cm. How long is the other leg, measured to the nearest hundredth?

   (A) 106.15 cm   (B) 30.00 cm   (C) 150.00 cm

   (D) 6.00 cm   (E) 36.00 cm

3. What is the volume of a rectangular prism with dimensions of 8 cm, 10 cm, and 4 cm?

   (A) 84 cm^3 (B) 320 cm^3 (C) 22 cm^3

   (D) 80 cm^3 (E) 304 cm^3

4. To plot the point $(-4, 3)$ in the coordinate plane, start at the origin and move
   (A) 4 units to the right and up 3 units.
   (B) 4 units to the left and up 3 units.
   (C) 4 units to the right and down 3 units.
   (D) 4 units to the left and down 3 units.
   (E) 4 units down and 3 units to the right.

5. What is the midpoint of the line segment with endpoints $(-5, -6)$ and $(1, 2)$?
   (A) $(-2, -2)$ (B) $(-2, -4)$ (C) $(-4, -4)$
   (D) $(-4, -2)$ (E) $(6, 8)$

6. The distance between the points $(1, 3)$ and $(7, 11)$ is equal to how many units?
   (A) 10   (B) 14   (C) $10\sqrt{10}$ (D) 100   (E) $8\sqrt{6}$

7. What is the slope of the line between the points $(6, 2)$ and $(3, -2)$?
   (A) 4   (B) $\dfrac{4}{3}$ (C) $\dfrac{3}{4}$ (D) 0   (E) 3

8. Which of the following represents the y-intercept of the linear equation $y + 3 = 2x$?
   (A) $\dfrac{1}{2}$ (B) 2   (C) -2 (D) 3   (E) -3

9. What is the solution to the system of equations shown in the graph below?

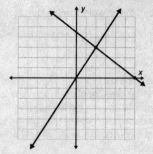

(A) $(2,3)$ (B) $(3,2)$ (C) $(2,0)$ (D) $(0,5)$ (E) $(6,0)$

10. Which answer choice best describes the transformation in the figure below?

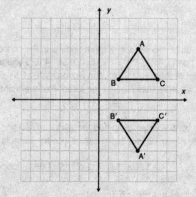

(A) reflection over the *y*-axis

(B) rotation of 90 degrees

(C) translation of $(+5, -2)$

(D) rotation of 180 degrees

(E) reflection over the *x*-axis

Answers and Explanations

1. C

This question reviews the area of a triangle formula. Use the formula $Area = \frac{1}{2} \times base \times height.$ $A = \frac{1}{2} \times 6 \times 8 = \frac{1}{2} \times 48 = 24$ square inches.

2. B

This question reviews the concept of Pythagorean theorem, or $a^2 + b^2 = c^2$. Since the measure of the hypotenuse is 78 cm and the measure of one leg is 72 cm, use the formula $a^2 + 72^2 = 78^2$. Evaluate the exponents: $a^2 + 5,184 = 6,084$. Then, subtract 5,184 from each side of the equation: $a^2 = 900$. Finally, solve: $a = \sqrt{900} = 30.00$ cm.

3. B

This questions reviews the concept of volume. Substitute into the volume formula $V = lwh$, where V is the volume, and l, w, and h represent the dimensions of the prism. The formula becomes $V = 8 \times 10 \times 4 = 320$ cm^3.

4. B

This problem tests your skills in graphing points in the coordinate plane. In the point (-4, 3), the x-coordinate of −4 tells you to move 4 units to the left of the origin. Then, the y-coordinate of 3 tells you to move 3 units up from there to find the location of the point.

5. A

This question tests your ability to find the midpoint of a line segment in the coordinate plane. To find the midpoint, use the formula $\left(\frac{x_1 + x_2}{2}, \frac{y_1 + y_2}{2}\right)$ and substitute the values of each endpoint. $\left(\frac{-5 + 1}{2}, \frac{-6 + 2}{2}\right) = \left(\frac{-4}{2}, \frac{-4}{2}\right) = (-2, -2)$.

6. A

This question addresses the concept of distance on a coordinate grid. To find the distance, use the formula $d = \sqrt{(x_1 - x_2)^2 + (y_1 - y_2)^2}$ and substitute the values of each point. $d = \sqrt{(1-7)^2 + (3-11)^2} =$

$\sqrt{(-6)^2 + (-8)^2} = \sqrt{36 + 64} = \sqrt{100} = 10$ units.

7. B

This question tests the concept of slope. To find the slope, use the

formula $m = \dfrac{\text{change in } y}{\text{change in } x} = \dfrac{y_1 - y_2}{x_1 - x_2}$ and substitute the values

of each point. $m = \dfrac{2 - (-2)}{6 - 3} = \dfrac{2 + 2}{3} = \dfrac{4}{3}$.

8. E

The problem in this question assesses the concept of the slope-intercept form of an equation. First, convert the equation to $y = mx + b$ form by subtracting 3 from each side. The equation becomes $y = 2x - 3$. Since the value of b in the equation is -3, the y-intercept is -3.

9. A

This question asks you to find the solution to a system of equations using a coordinate grid. Since the two linear equations in the graph intersect at the point (2, 3), this point is the solution to the system of equations.

10. E

Question 10 assesses your knowledge of transformations in a coordinate system. Triangle ABC is flipped over the x-axis to become triangle A'B'C'. The image of triangle ABC is inverted and appears to be a mirror image of the original triangle. This is best described as a reflection over the x-axis.

COORDINATE GEOMETRY

Have you ever tried to locate a city on a map using a key, or used a grid to help you calculate an area? Coordinate geometry is a way to use a rectangular grid to locate particular places and determine measurements such as area or distance. In this chapter you will learn about the coodinate plane, along with various applications such as the midpoint formula, distance formula, slope, linear equations and inequalities, and systems of equations. In addition, you will be introduced to transformational geometry.

The Coordinate Plane

The coordinate plane is formed by the intersection of two perpendicular number lines. The horizontal number line is known as the x-axis and the vertical number line is known as the y-axis. When the lines intersect, four regions, called **quadrants**, are formed. They are numbered I, II, III, IV, in counter-clockwise fashion, starting from the upper right-hand quadrant. The point where the two number lines intersect is called the **origin**, and has the coordinates (0, 0). The number lines are labeled with positive numbers to the right of the origin on the x-axis and above the origin on the y-axis, and with negative numbers to the left of the origin on the x-axis and below the origin on the y-axis.

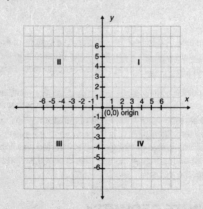

Each point in the coordinate system has a location determined by the number of spaces the point lies to the right or left of the origin, and the number of spaces it lies above or below the origin. Therefore, each point in the system is named with two numbers: an x-coordinate and a y-coordinate. These coordinates are always written in x,y order and placed in parentheses. Take, for example, the point named by the coordinates (4, 5). The first number in the pair is 4, so this point is 4 spaces to the right of the origin on the *x*-axis. The second number is 5, so the point is also 5 spaces above the origin on the *y*-axis. Therefore, to find the location of this point, start at the origin, move 4 spaces to the right, and 5 spaces up from there. Note the location in quadrant I of the point (4, 5) in the figure below.

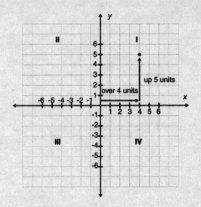

MATH SPEAK

Each point (*x*, *y*) in the coordinate system has an *x*-value, also known as the **abscissa**, and a *y*-value, also known as the **ordinate**. Every point in the coordinate plane can be named using these two values.

Here are some additional examples of how to find the locations of other points in the coordinate plane. The location of each is noted in the figure on the following page.

Point A: To find the location of point A, start at the origin, move 2 spaces to the left on the *x*-axis, and then move 6 spaces up. This is the point (–2, 6), located in quadrant II.

Point B: To find the location of point B, start at the origin, move 3 spaces to the left on the x-axis, and then move 4 spaces down. This is the point (−3, −4), located in quadrant III.

Point C: To find the location of point C, start at the origin, move 5 spaces to the right on the x-axis, and then move 6 spaces down. This is the point (5, −6), located in quadrant IV.

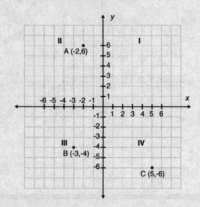

Notice the following pattern when graphing points in the four quadrants:

In quadrant I, both the x and y coordinates are positive.

In quadrant II, all x-values are negative while the y-values are positive.

In quadrant III, both the x and y coordinates are negative.

In quadrant IV, all x-values are positive, while the y-values are negative.

Repetition is important if you want to master plotting points in the coordinate plane . Try the next set of practice questions to help develop proficiency in identifying coordinates.

Practice 1

Use the graph below to match each point with the correct coordinates. Answers and explanations are located at the end of the chapter.

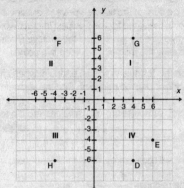

1. Point D
2. Point E
3. Point F
4. Point G
5. Point H

A. $(-4, 6)$
B. $(4, -6)$
C. $(-4, -6)$
D. $(6, -4)$
E. $(4, 6)$

Formulas

The Midpoint Formula

It is often helpful to know the halfway point, or **midpoint**, between two endpoints of a line segment. You can find this location by

using the formula $\left(\dfrac{x_1 + x_2}{2}, \dfrac{y_1 + y_2}{2}\right)$, where (x_1, y_1) and (x_2, y_2)

represent the two endpoints. To find the midpoint between the points $(2, 5)$ and $(4, -3)$, for example, first plug in the values of x and y from

each point and then evaluate the formula. $\left(\dfrac{x_1 + x_2}{2}, \dfrac{y_1 + y_2}{2}\right) =$

$\left(\dfrac{2+4}{2}, \dfrac{5+-3}{2}\right) = \left(\dfrac{6}{2}, \dfrac{2}{2}\right) = (3, 1)$ The midpoint between the points (2, 5) and (4, -3) is (3, 1).

To help you remember the midpoint formula, notice that you are actually finding the sum of the two x-values and dividing it by 2, and then you are doing the same for the y-values. In other words, you are finding the average (mean) x-value and the average (mean) y-value when you are finding the midpoint between two points.

You can also use this formula to find an endpoint when given the value of the other endpoint and the midpoint. Take the segment $\overline{AB}$ with point A at (4, -7) and the midpoint of $\overline{AB}$ at (-1, -3). To find the location of point B, substitute the known values into the formula and set it equal to the midpoint: $\left(\dfrac{x_1+x_2}{2}, \dfrac{y_1+y_2}{2}\right) = \left(\dfrac{4+x_2}{2}, \dfrac{-7+y_2}{2}\right)$ = (-1, -3) Now, set each expression in the formula equal to its coordinate at the midpoint: $\dfrac{4+x_2}{2} = -1$ and $\dfrac{-7+y_2}{2} = -3$.

Cross-multiply in the first equation: $4 + x_2 = -2$, so $x_2 = -6$. Cross-multiply in the second equation: $-7 + y_2 = -6$, so $y_2 = 1$. Therefore, the coordinates of point B are (-6, 1).

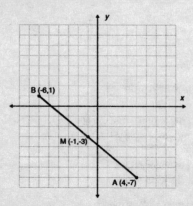

The Distance Formula

Another formula frequently used in the coordinate plane is the **distance formula**. This formula will help you calculate the distance between any two points in the coordinate plane. To find the distance between any two points (x_1, y_1) and (x_2, y_2), use the formula $d = \sqrt{(x_1 - x_2)^2 + (y_1 - y_2)^2}$, which is based on the Pythagorean theorem. For example, to find the distance between the points $(0, -2)$ and $(5, -2)$, substitute the values for x and y from each point into the formula:

$$\sqrt{(x_1 - x_2)^2 + (y_1 - y_2)^2} = \sqrt{(0 - 5)^2 + (-2 - (-2))^2} =$$
$$\sqrt{(-5)^2 + (0)^2} = \sqrt{25 + 0} = \sqrt{25} = 5.$$

The distance between the points $(0, -2)$ and $(5, -2)$ is 5 units.

The distance between two points may also be an irrational number.

FLASHBACK

Review information regarding **irrational numbers** from chapter 1 and **simplifying square roots** from chapter 7.

Take the points $(4, 2)$ and $(-3, 6)$. To find the distance between them, substitute into the distance formula:

$$\sqrt{(x_1 - x_2)^2 + (y_1 - y_2)^2} = \sqrt{(4 - (-3))^2 + (2 - 6)^2} =$$
$$\sqrt{(7)^2 + (-4)^2} = \sqrt{49 + 16} = \sqrt{65}.$$

Notice in this case the final distance is not a perfect square. Since the square root of 65 is approximately equal to 8, the distance can be rounded to 8 units using a calculator, but the exact answer is $\sqrt{65}$ units.

> **REMEMBER THIS!**
>
> The **midpoint** between two points can be found using the formula
>
> $$\left(\frac{x_1 + x_2}{2}, \frac{y_1 + y_2}{2}\right).$$
>
> The **distance** between two points can be found using the formula
>
> $$d = \sqrt{(x_1 - x_2)^2 + (y_1 - y_2)^2}.$$

Practice 2

Use the following bank of answers to correctly fill in each blank below. Answers and explanations are located at the end of the chapter.

$(-11, -12)$ $(-2, -5)$ $(-1, -5)$ $(11, -12)$ $\sqrt{13}$ 13 $3\sqrt{3}$ $\sqrt{29}$ 29

6. The midpoint between $(2, -3)$ and $(-4, -7)$ is _____.

7. If point C is $(5, 4)$ and the midpoint of segment CD is $(-3, -4)$, then point D is located at _____.

8. The distance between $(-2, -3)$ and $(-7, -15)$ is _____.

9. The distance between the points $(1, 4)$ and $(6, 2)$ is _____.

Slope and Its Applications

Slope is an important concept to master when studying coordinate geometry. It helps you draw conclusions about the pattern of a graph, and tells you the *rate of change* of a situation.

MATH SPEAK

The slope (m) of the line between two points (x_1, y_1) and (x_2, y_2) can be found by using the formula

$$m = \frac{\text{change in } y}{\text{change in } x} = \frac{y_1 - y_2}{x_1 - x_2}.$$

Slope is commonly known as the *rise over the run*. In other words, the number in the numerator of the fraction tells how many units to move up or down and the number in the denominator tells how many units to move to the right or left. If the slope is written as a whole number, write that value over the number 1. A slope of 3 is written as $\frac{3}{1}$ because the rise is 3 and the run is 1.

Read the following example to see the slope formula in action:

Find the slope of the line between the points (4, 8) and (-1, 6).

Use the formula $m = \dfrac{\text{change in } y}{\text{change in } x} = \dfrac{y_1 - y_2}{x_1 - x_2} = \dfrac{8 - 6}{4 - (-1)}$

$= \dfrac{2}{5}$. The slope of the line between these points is $\dfrac{2}{5}$. The rate of change between the points on this line is up 2 units, and then over 5 units.

Another type of slope problem is one where the slope of the line is given, but one or more of the coordinates of a point(s) on the line is unknown.

Find the value of t if the slope of the line between the points (1, t) and (2, 1) is 2.

Use the slope formula with the given coordinates and set it equal to the slope of 2, or $\dfrac{2}{1}$. $\dfrac{y_1 - y_2}{x_1 - x_2} = \dfrac{t - 1}{1 - 2} = \dfrac{2}{1}$. Simplify the left side of the proportion: $\dfrac{t - 1}{-1} = \dfrac{2}{1}$. Cross-multiply to get $t - 1 = -2$. Add 1 to both sides of the equation to get $t = -1$. The missing y-coordinate is -1.

Special Cases of Slope

There are four major categories that the slope of a line can fall under: positive slope, negative slope, slope of 0, or no slope. *Positive slopes* are lines that go up to the right as you move from left to right. When using the slope, count up and over to the right when plotting the graph. These lines appear to be "uphill," as shown in the figure below.

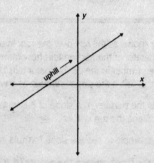

Negative slopes are lines that go up to the left, and appear to be at a downhill slant. With negative slope, count up and over to the left when plotting the graph. The figure below shows a line with negative slope.

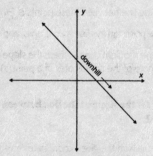

Two additional special cases of slope are horizontal and vertical lines. When the rate of change in the *y*-values is equal to 0, the line formed will be a horizontal line. In this case, the slope would have a rise of 0 and a run of any real number. This would cause the graph to travel horizontally across, but to not have any steepness. Each horizontal line is in the form $y = k$, where k represents a constant value. A few examples are shown in the figure below.

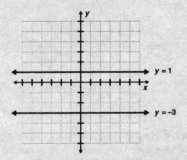

When the rate of change of the x-values is 0, then the line formed is a vertical line. Because the 0 will then be at the bottom of your fraction, no real number slope can exist. Each vertical line is in the form $x = k$, where k represents a constant value. A few examples of vertical lines and their equations are shown in the figure below.

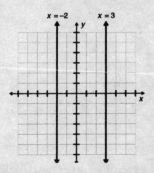

Intercepts

There are two intercepts that are commonly used when graphing lines in the coordinate plane. The first is the x-intercept. The x-intercept is the location at which a line intersects, or crosses, the x-axis. This point is written in the form $(x, 0)$; x is the point where the line crosses the x-axis. Notice that the y-value at the x-intercept will always be 0.

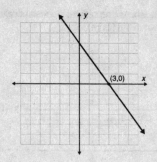

The y-intercept is the location at which a line intersects the y-axis. This point is written in the form (0, y); y is the point where the line crosses the y-axis. Also notice in this case the x-value at the y-intercept will always be 0.

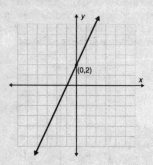

Practice 3

Answer true or false for each of the following questions.

10. **T   F**   The slope of a line can be found by calculating the change in x over the change in y.

11. **T   F**   The slope of the line between (6, –1) and (–1, 3) is $-\frac{4}{7}$.

12. **T**   **F**   The line in the form $y = 5$ is a horizontal line.

13. **T**   **F**   The line in the form $x = -3$ is a horizontal line.

14. **T**   **F**   The x-intercept of a line could be at the point
$(0, -2)$.

15. **T**   **F**   The y-intercept of a line could be at the point
$(0, -10)$.

Graphing in Slope-Intercept Form

Now that you have learned about slopes and intercepts in detail, you will have a better understanding of how to graph lines and use slope-intercept form.

> ### MATH SPEAK
>
> The **slope-intercept form** of a linear equation is $y = mx + b$, where m is the slope of the line and b is the y-intercept.

In order to graph a line written in slope-intercept form, follow these four steps.

1) Make sure the equation is in the form $y = mx + b$.

2) Use b from the equation (the y-intercept) to graph the point $(0, b)$ on the y-axis.

3) Using the numerator and denominator of m, or the slope of the equation, start at the y-intercept and count up (the amount of the rise) and over (the amount of the run) to find another point. Repeat the process of counting up and over to find additional points on the line.

4) Connect the points to create your line.

Practice this procedure with the linear equation $y = 3x + 1$.

1) Since the equation is in the form $y = mx + b$, first identify that $m = 3$ and $b = 1$.

2) Place a point at the y-intercept of $(0, 1)$.

3) Since the slope is 3 or $\frac{3}{1}$, count up 3 units and over 1 unit to find another point on the line. This is the point (1, 4). Note that since the slope is positive, count up and over to the right. Repeat the process of counting up 3 and over 1 to the right to find additional points.

4) Connect the points to form the graph of the line.

This equation is graphed in the figure below.

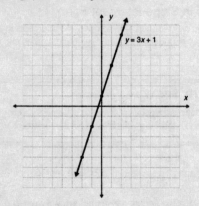

$y = 3x + 1$

Try another practice example using the equation $2y + x = 4$.

1) Since the equation is not in the form $y = mx + b$, first subtract x from both sides of the equation and then divide each term by 2 to get

$y = -\frac{1}{2}x + 2$. Now, identify that $m = -\frac{1}{2}$ and $b = 2$.

2) Place a point at the y-intercept of (0, 2).

3) Since the slope is $-\frac{1}{2}$, count up 1 unit and over 2 units to the left to find another point on the line. This is the point (−2, 3). Note that since the slope is negative, you are counting up and over to the left. Repeat the process of counting up 1 and over 2 to the left to find additional points.

4) Connect the points to form the graph of the line.

This equation is graphed in the figure below.

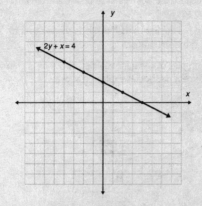

You can use linear equations to represent real-world situations. For example, say a health club charges an initial $50 to register, and then $25 per month. This can be modeled with the equation $y = 25x + 50$, where x is the number of months and y is the total cost of the membership. The slope of the equation is the rate of change; in other words, the monthly fee. The y-intercept of the equation is the initial, or one-time, cost. Using this equation, you can determine the cost of joining this club for one year. Since there are 12 months in a year, substitute $x = 12$ into the equation.

$$y = 25(12) + 50$$
$$y = 300 + 50$$
$$y = 350$$

The membership would cost $350 for 1 year.

Linear Inequalities

Inequalities use the symbols $<$ (less than), $>$ (greater than), $\leq$ (less than or equal to), and $\geq$ (greater than or equal to) instead of equal signs. Linear inequalities work just like linear equations, except for 2 differences:

1. The line should be *dashed*, not solid, if the inequality symbol is < or > because the points that lie on the line are not in the solution set.

2. One side of the line, or half-plane, is shaded to show all the points in the solution set.

Below are examples of linear inequalities. Each one is graphed using the slope-intercept method as explained above.

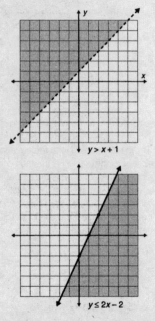

$y > x + 1$

$y \leq 2x - 2$

Note that in the figure above, the line for the inequality $y > x + 1$ is dashed because the symbol is *greater than* and not *greater than or equal to*. The *greater than* symbol indicates that the graph should be shaded above the line.

In the other figure above, the line for the inequality is solid because the symbol is *less than or equal to*. This symbol also causes the shading to fall below the line.

> ## REMEMBER THIS!
>
> When graphing a linear inequality that contains < or >, make the line dashed.
>
> If the inequality is in slope-intercept, shade below the line for < or ≤ and above the line for > or ≥.

Practice 4

Match the sketch of the graph with the correct equation. Answers and explanations are located at the end of the chapter.

16. $y = 2x + 1$ 17. $y = -2x + 1$ 18. $y = 2x - 1$

19. $y < x - 1$ 20. $y \geq x + 1$

A.

B.

C.

D.

E.

Fill in the blank to make each statement true.

21. If a video at Video Valley costs $2.50 to rent and the initial membership fee is $4, this can be represented by the linear equation _____.

22. If the cost of a cab ride can be represented by the linear equation $y = 2x + 3$, where x represents the number of miles and y is the total cost, the cost per mile is _____ dollars.

Solving Systems of Equations

When more than one equation is graphed on the same set of axes, a system of equations is created. To solve a system of equations, look for the intersection of the lines. These point(s) of intersection, if they exist, are the solution(s) to the system.

Here is an example of how to solve a system of equations. Take the system:

$$y = -2x + 3$$
$$y = x - 3$$

First, graph each equation on the same set of axes. In the first equation, the slope is equal to -2 and the y-intercept is 3. In the second equation, the slope is 1 and the y-intercept is -3. The figure below shows these two equations graphed on the same set of axes.

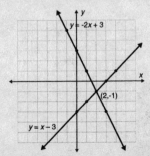

Since the two lines intersect at the point $(2, -1)$, this point is the solution to the system of equations. To check this solution, substitute the coordinates $x = 2$ and $y = -1$ into each of the original equations and check to make sure they are equal.

$$y = -2x + 3 \qquad\qquad y = x - 3$$
$$-1 = -2(2) + 3 \qquad -1 = 2 - 3$$
$$-1 = -4 + 3 \qquad\qquad -1 = -1$$
$$-1 = -1$$

Here is another example of a system of equations:

$$y + x = 4$$
$$y = x - 2$$

Again, first graph each equation on the same set of axes using the slope-intercept form. Since the first equation is not in slope-intercept form, subtract x from each side of the equation to get $y = -x + 4$. The slope of this line is -1 and the y-intercept is 4. In the second equation, the slope is 1 and the y-intercept is -2. The figure below shows these two lines graphed on the same set of axes, intersecting at the point (3, 1).

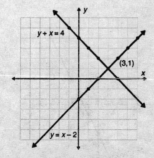

What is special about the two lines in this system is that they meet to form 90°, or right angles. In other words, these two lines are perpendicular. Notice that the slope of the first line was -1 and the slope of the second line was 1. If the slopes of two lines are *negative reciprocals* like -1 and 1, or -2 and $\frac{1}{2}$, then the lines are perpendicular.

FLASHBACK

To review different types of lines, including **perpendicular lines**, refer to the geometry concepts in chapter 7.

Special Systems of Equations

There are three different cases of solutions to systems of equations when working with linear equations: one solution, infinite solutions, or no solution.

The first case was discussed above when one solution was found for the system—the two lines intersect at a single point.

The second case is when the lines appear to have different equations, but end up being the same line after the equations are transformed to $y = mx + b$ form. These lines are called *coincident lines* and actually share all the same points. Therefore, there are infinite solutions to a system of coincident lines.

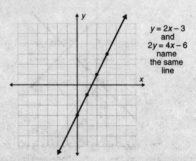

$y = 2x - 3$
and
$2y = 4x - 6$
name
the same
line

The third case is when the lines do not intersect at all. The only way this can happen in a plane is if the lines are parallel. Parallel lines in the same plane slant at the same rate, running next to each other but never touching. Therefore, parallel lines have the same slope. If two lines on the same set of axes have the same slope, then there is no solution to this system of parallel lines. In the figure below, each line has a slope of -1; therefore the lines are parallel and will not intersect.

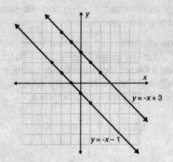

REMEMBER THIS!

Parallel lines have the same slope.

Perpendicular lines have slopes that are negative reciprocals of one another.

Practice 5

Fill in each blank with the best possible answer. Answers and explanations are located at the end of the chapter.

23. The solution to the system $y = 2x - 1$ and $y = -x + 5$ is

_____.

24. The system of equations $y = 2x - 5$ and $-4x + 2y = -10$ has _____ solution(s).

25. In the system $6y = 3x + 6$ and $y = \frac{1}{2}x - 1$, there is (are) _____ solution(s).

26. _____ lines have the same slope.

27. _____ lines have slopes that are negative reciprocals.

Transformations in the Coordinate Plane

When something is transformed it is changed in some way. There are four basic types of transformations in the coordinate plane: reflections, translations, rotations, and dilations.

The first type of transformation is a **reflection**. When dealing with reflections, think of a mirror image. A reflection is also known as a "flip," as the object being reflected appears to flip over a line of reflection. Take the example below. Triangle ABC is reflected, or flipped, over the y-axis and triangle A'B'C' is the image of reflection.

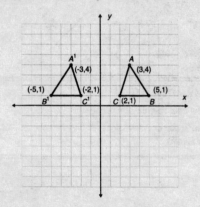

This reflection is written as $\triangle ABC \xrightarrow{Ry\text{-}axis} \triangle A'B'C'$, and the line of reflection is the y-axis. Notice that the y-coordinates remain the same, while the x-coordinates become their opposites. In the same manner, if reflecting over the x-axis, the x-coordinates would remain the same while the y-coordinates would become their opposites.

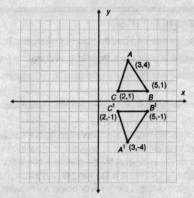

Another common line of reflection is the line $y = x$. In this linear equation, the slope of the line is 1 (the coefficient of x is 1) and the y-intercept is 0 (there is no b in this equation). This line contains all of the points where x is equal to y, such as $(-1, -1)$, $(0, 0)$, $(2, 2)$, etc. When reflecting over this line, all of the coordinates of the points in the object are reversed; in other words, (x, y) is changed to (y, x). An example of this is shown in the figure below.

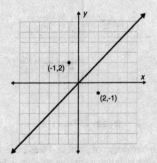

> ### REMEMBER THIS!
>
> In a reflection over the x-axis, the point (x, y) changes to $(x, -y)$.
> In a reflection over the y-axis, the point (x, y) changes to $(-x, y)$.
> In a reflection over the line $y = x$, the point (x, y) changes to (y, x).

The next type of transformation is a **translation**. With a translation, an object is moved a certain number of units either left or right and then either up or down, depending on the translation. A translation is written using the notation $T_{a, b}$ where a is the number of horizontal units (change in the x-coordinates) the point or object will move and b is the number of vertical units (change in the y-coordinates) the point or object will move. If the value of a or b is positive, that number is added to the corresponding coordinates; if the value of either is negative, it is subtracted from the corresponding coordinates. Take the example below.

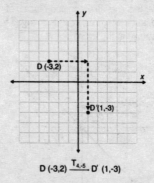

$$D \ (-3,2) \xrightarrow{T_{4,-5}} D' \ (1,-3)$$

Notice in the above translation of point D, 4 was added to each x-coordinate and -5 was added to each y-coordinate to find the image.

> ### REMEMBER THIS!
>
> Any translation $T_{a, b}$ transforms the point (x, y) to $(x + a, y + b)$.

The third type of transformation is a **rotation**. A rotation is also known as a "turn," as the object is turned much like a wheel would spin about a center point. The notation *Rot*, or sometimes just *r*, is used to indicate a rotation. The figure below shows a few of the most common types of rotations, each one with the center at the origin.

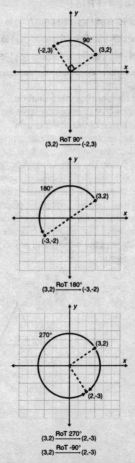

Note that any positive rotation is in a counter-clockwise fashion, and any negative rotation occurs in a clockwise direction. As shown in the figure above, a rotation of 270° is the same as a rotation of −90°.

REMEMBER THIS!

A rotation of 90° transforms the point (x, y) to $(-y, x)$.

A rotation of 180° transforms the point (x, y) to $(-x, -y)$.

A rotation of −90° transforms the point (x, y) to $(y, -x)$.

The fourth type of transformation is a **dilation**. The dilation is the only type of transformation that does not preserve the size of the object. In most dilations, the image of the object is smaller or larger than the original object, but always remains in proportion to the original object. Each dilation has a center (which is usually at the origin in a coordinate plane) and a scale factor. The scale factor indicates how many times larger or smaller the image is of the original figure. In other words, the scale factor is multiplied by the coordinates of the original image to find the coordinates of the new image. The graph below shows a dilation of scale factor 2. Notice how each coordinate is multiplied by 2 to find the coordinates of the image. Each dimension of the dilated image is twice as large as the original and is also twice as far away from the center, or the origin.

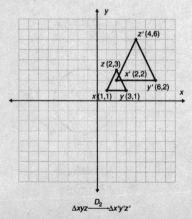

$$\Delta xyz \xrightarrow{\;D_2\;} \Delta x'y'z'$$

> ## REMEMBER THIS!
>
> A dilation D with scale factor c changes the point (x, y) to $(c \times x, c \times y)$.

Transformations in the coordinate plane can also occur with graphs of algebraic equations, not just geometric figures. For example, take the equation $y = x^2$. The graph of this function is a parabola, or u-shape, as shown below.

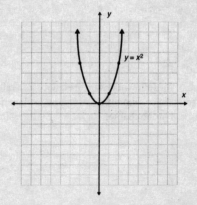

$y = x^2$

To find the graph of the equation $y = x^2 + 2$, this graph would be the same as $y = x^2$ just moved up, or translated, 2 units. Notice that the $+2$ at the end of the equation changes the y-intercept from 0 to 2, but the rest of the figure remains the same shape. The translation is shown on the following page.

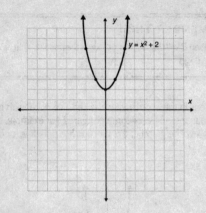

$$y = x^2 + 2$$

Practice Set 6

Describe each of the following transformations as a reflection, translation, rotation or dilation. Answers and explanations are located at the end of the chapter.

28.

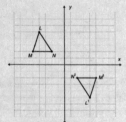

29.

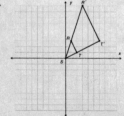

30.

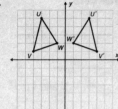

31.

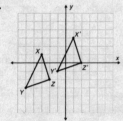

SUMMARY

Coordinate geometry is an important application in mathematics that involves both algebraic and geometric concepts. In conclusion, remember these five things about working in the coordinate plane:

- When plotting points, move left or right the number of units of the x-coordinate and move up or down the number of units of the y-coordinate.

- Use the midpoint, distance, and slope formulas for calculating these measures. Use the slope-intercept method ($y = mx + b$) to graph linear equations.

- To solve a system of equations, look for the point of intersection of the lines. A linear system of equations will have one solution, infinite solutions, or no solution.

- Parallel lines have the same slope and perpendicular lines have slopes that are negative reciprocals of one another.

- The four basic types of transformations in the coordinate plane are reflections, translations, rotations, and dilations.

Practice Answers and Explanations

1.   B

This point is located in quadrant IV, 4 units to the right and 6 units below the origin.

2.   D

This point is located in quadrant IV, 6 units to the right and 4 units below the origin.

3.   A

This point is located in quadrant II, 4 units to the left and 6 units above the origin.

4.   E

This point is located in quadrant I, 4 units to the right and 6 units above the origin.

5.   C

This point is located in quadrant III, 4 units to the left and 6 units below the origin.

6.   (–1, –5)

The midpoint can be found by substituting the points into the formula

$$\left(\frac{x_1 + x_2}{2}, \frac{y_1 + y_2}{2}\right) = \left(\frac{2 + -4}{2}, \frac{-3 + -7}{2}\right) = \left(\frac{-2}{2}, \frac{-10}{2}\right) = (-1, -5).$$

7.   (–11, –12)

The point can be found by substituting into the midpoint formula and setting it equal to the known midpoint.

$$\left(\frac{x_1 + x_2}{2}, \frac{y_1 + y_2}{2}\right) = \left(\frac{5 + x_2}{2}, \frac{4 + y_2}{2}\right) = (-3, -4).$$ Now, set each

expression in the formula equal to its coordinate at the midpoint.

$$\frac{5 + x_2}{2} = -3$$ and $$= \frac{4 + y_2}{2} -4.$$ Cross-multiply in the first equation to

get $5 + x_2 = -6$, so $x_2 = -11$. Cross-multiply in the second equation to get $4 + y_2 = -8$, so $y_2 = -12$. Therefore, the coordinates of point B are (–11, –12).

8. 13

Substitute the coordinates of each point into the distance formula.

$$\sqrt{(x_1 - x_2)^2 + (y_1 - y_2)^2} = \sqrt{((-2) - (-7))^2 + ((-3) - (-15))^2} =$$
$$\sqrt{(5)^2 + (12)^2} = \sqrt{25 + 144} = \sqrt{169} = 13.$$

9. $\sqrt{29}$

Substitute the coordinates of each point into the distance formula.

$$\sqrt{(x_1 - x_2)^2 + (y_1 - y_2)^2} = \sqrt{(1 - 6)^2 + (4 - 2)^2} = \sqrt{(-5)^2 + (2)^2}$$
$$= \sqrt{25 + 4} = \sqrt{29}.$$

10. False

The slope of a line is the change in y over the change in x.

11. True

Use the slope formula. $m = \dfrac{\text{change in } y}{\text{change in } x} = \dfrac{y_1 - y_2}{x_1 - x_2} =$
$\dfrac{-1 - 3}{6 - (-1)} = \dfrac{-4}{7}$.

12. True

Any line with the equation $y = k$ is a horizontal line.

13. False

Any line with the equation $x = k$ is a vertical line.

14. False

The point $(0, -2)$ lies on the y-axis. This point could be a y-intercept, not an x-intercept.

15. True

The point $(0, -10)$ lies on the y-axis and could be the y-intercept of an equation.

16. C

In this figure, the line has slope of 2 and the y-intercept is 1.

17. A

In this figure, the line has slope of -2 and the y-intercept is 1.

18. D

In this figure, the line has slope of 2 and the y-intercept is −1.

19. E

In this figure, the dashed line and shading below the line indicate a *less than* inequality. In addition, the line has slope of 1 and a y-intercept of −1.

20. B

The shading above the solid line indicates a *greater than or equal to* inequality. In addition, the line has slope of 1 and a y-intercept of 1.

21. $y = 2.5x + 4$

Since each video costs $2.50 to rent, then $m = 2.50$ or 2.5. The initial cost of the membership is $4, so this is the y-intercept, or b.

22. $2 per mile

The cost per mile will be the value of m in the linear equation, which is 2.

23. (2, 3)

When the two lines are graphed on the same set of axes, they cross at the point (2, 3). This is shown in the figure below.

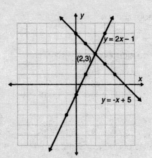

24. infinite

When the equation $-4x + 2y = -10$ is converted to $y = mx + b$ form, it becomes $y = 2x − 5$. Since the equations are the same, the lines coincide and every point on the line is in the solution set. Therefore, there are infinite solutions.

25. no

When the equations are in $y = mx + b$ form, they each have slope of $\frac{1}{2}$. Since they have the same slope and are not the same line, the lines are parallel and will never intersect. Therefore, there is no solution to this system of equations.

26. parallel

Parallel lines have the same slope and will never intersect.

27. perpendicular

Perpendicular lines meet to form right angles and their slopes are negative reciprocals.

28. rotation

The figure is rotated, or turned, about the origin 180 degrees.

29. dilation

Triangle RST is enlarged, or dilated, to becomes triangle R'S'T'.

30. reflection

Triangle U'V'W' is a mirror image, or reflection, of triangle UVW.

31. translation

A translation, or slide, moves triangle ABC to the location of triangle A'B'C'.

CHAPTER 10 TEST

Try the following questions to test your knowledge of coordinate geometry.

1. In the figure below, what letter corresponds with the coordinates $(4, -2)$?

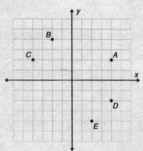

(A) A (B) B (C) C (D) D (E) E

2. In which quadrant is the point $(-2, 3)$ located?

(A) I (B) II (C) III (D) IV (E) none of these

3. If the endpoints of segment AB are $A(3, 8)$ and $B(-1, 2)$, what is the midpoint of segment AB?

(A) $(-2, 5)$ (B) $(2, 5)$ (C) $(1, 2)$ (D) $(5, 1)$ (E) $(1, 5)$

4. If the midpoint M of line segment CD has coordinates $(6, -2)$ and C is $(3, 4)$, what are the coordinates of point D?

(A) $(-8, -9)$ (B) $(9, 2)$ (C) $(9, -8)$
(D) $(3, 1)$ (E) $(-3, -1)$

5. Which of the following represents the distance between the points $(-2, 5)$ and $(3, 0)$?

(A) 5 (B) 25 (C) 50 (D) $10\sqrt{2}$ (E) $5\sqrt{2}$

6. If a line containing the two points $(7, 3)$ and $(t, 5)$ has slope of $\frac{2}{3}$, what is the value of t?

   (A) –10   (B) 6   (C) 20   (D) 10   (E) –6

7. What is the slope of the line $2x - y = 5$?

   (A) –2   (B) 2   (C) 5   (D) –5   (E) $\frac{5}{2}$

8. Which of the following lines has slope equal to 0?
   (A) $y = 3x + 1$ (B) $y = x$ (C) $y = 4$
   (D) $x = 0$ (E) $x = -3$

9. Which of the following equations could represent the line in the figure below?

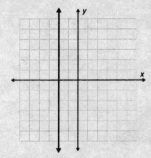

   (A) $x = 2$ (B) $x = -2$ (C) $x = 0$
   (D) $y = 2$ (E) $y = -2$

10. Which of the following answer choices represents the
inequality $y < x + 3$?

(A)

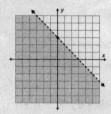

(B)

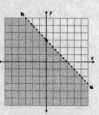

(C)

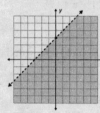

(D)

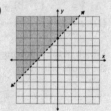

(E)

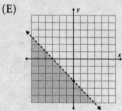

11. Based on the graph below, which of the following is the solution to the system of equations?

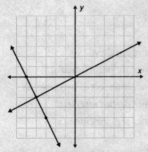

   (A) $(-2, 4)$ (B) $(-4, 2)$ (C) $(-6, 0)$
   (D) $(-4, -2)$ (E) $(0, -6)$

12. What is the solution to the following system of equations?

   $y = 3x - 1$
   $-3x + y = 5$

   (A) $(1, 5)$ (B) $(0, -1)$ (C) $(-3, 1)$
   (D) $(-3, 5)$ (E) There is no solution.

13. Which of the following equations would NOT intersect with the line $2y + 4 = x$?

   (A) $y = 2x + 4$ (B) $y = -2x + 2$ (C) $y = \frac{1}{2}x + 4$
   (D) $y = -\frac{1}{2}x + 4$ (E) $y = -2x - 2$

14. Which of the following statements best describes the system of equations below?

$$3y = 3x + 9$$
$$y - x = 3$$

(A) The lines are skew.

(B) The lines are perpendicular.

(C) The lines intersect at two points.

(D) The lines do not intersect.

(E) The lines are coincident.

15. The cost of parking at a city lot is $5.50 for the first hour and $4.00 for each additional hour. If y represents the total cost and x is the number of additional hours parked, which of the following linear equations could represent this situation?

(A) $y = 4x + 5.50$ (B) $y = -4x + 5.50$ (C) $y = 4x - 5.50$

(D) $y = 5.5x + 4$ (E) $y = 5.5x - 4$

16. Using the situation in question 15, how much will it cost to park at this garage for 8 hours?

(A) $9.50   (B) $32.00   (C) $33.50

(D) $37.50   (E) $76.00

17. Which of the following represents a reflection over the line
$y = x$?

(A)

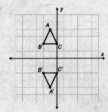

(B)

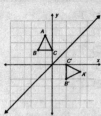

(C)

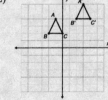

(D)

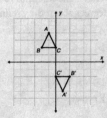

(E)

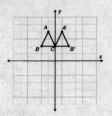

18. If the point $(3, -5)$ is moved to $(6, -2)$, the transformation is represented by which of the following translations?

   (A) $T_{9, -3}$

   (B) $T_{-3, 3}$

   (C) $T_{3, 3}$

   (D) $T_{3, -7}$

   (E) $T_{9, -7}$

19. What is the image of the point $(5, -6)$ after a rotation of 180 degrees about the origin?

   (A) $(-5, 6)$

   (B) $(-6, 5)$

   (C) $(6, -5)$

   (D) $(5, 6)$

   (E) $(-6, -5)$

20. If the graph below is the line $y = -x^2$, then which of the following is the graph of $y = -x^2 - 1$?

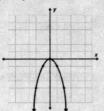

(A)

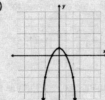

(B)

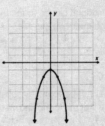

(C)

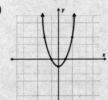

(D)

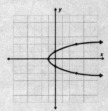

(E)

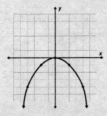

Answers and Explanations

1. D

To graph the point $(4, -2)$, start at the origin and move 4 units to the right. From there, move 2 units down. The letter at this placement is D.

2. B

To graph the point $(-2, 3)$, start at the origin and move 2 units to the left. From there, move 3 units up. This location is in the second quadrant, or quadrant II.

3. E

The midpoint can be found by substituting the coordinates for A and B into the midpoint formula.

$$\left(\frac{x_1 + x_2}{2}, \frac{y_1 + y_2}{2}\right) = \left(\frac{3 + -1}{2}, \frac{8 + 2}{2}\right) = \left(\frac{2}{2}, \frac{10}{2}\right) = (1, 5)$$

4. C

Point D can be found by substituting into the midpoint formula and setting it equal to the known midpoint.

$$\left(\frac{x_1 + x_2}{2}, \frac{y_1 + y_2}{2}\right) = \left(\frac{3 + x_2}{2}, \frac{4 + y_2}{2}\right) = (6, -2)$$

Now, set each expression in the formula equal to its coordinate at the midpoint. $\frac{3 + x_2}{2} = 6$ and $= \frac{4 + y_2}{2} -2$. Cross-multiply in the first equation to get $3 + x_2 = 12$, so $x_2 = 9$. Cross-multiply in the second equation to get $4 + y_2 = -4$, so $y_2 = -8$. Therefore, the coordinates of point D are $(9, -8)$.

5. E

Substitute the coordinates of each point into the distance formula.

$$\sqrt{(x_1 - x_2)^2 + (y_1 - y_2)^2} = \sqrt{((-2) - (3))^2 + (5 - 0)^2} =$$
$$\sqrt{(-5)^2 - (5)^2} = \sqrt{25 + 25} = \sqrt{50} = \sqrt{25}\sqrt{2} = 5\sqrt{2}$$

6.   D

Use the slope formula with the given coordinates and set it equal to the slope of $\frac{2}{3}$. $\frac{y_1 - y_2}{x_1 - x_2} = \frac{3 - 5}{7 - t} = \frac{2}{3}$. Simplify the left side of the proportion: $\frac{-2}{7 - t} = \frac{2}{3}$. Cross-multiply to get $14 - 2t = -6$. Subtract 14 from both sides of the equation, and then divide by -2 to get $t = 10$. The missing x-coordinate is 10.

7.   B

Convert the equation to $y = mx + b$ form. First, subtract $2x$ from each side of the equation to get $-y = -2x + 5$. Then, divide each side of the equation by -1 to get $y = 2x - 5$. The value of m, which is the slope, is 2.

8.   C

Any line in the form $y = k$ is a horizontal line and has no slope. The only line in this form is choice C. Choice A has slope of 3, choice B has slope of 1, and choices D and E are vertical lines that have undefined slope.

9.   B

Since the line in the figure is a vertical line, it is in the form $x = k$. The line crosses the x-axis at -2, so the equation of the line is $x = -2$.

10.   C

This inequality has a slope of 1 and a y-intercept of 3. Since the symbol used is *less than*, or $<$, the line should be dashed and the half-plane below the line is shaded. Each of these characteristics is true only for choice C.

11.   D

Since the two lines in the system of equations intersect at the point $(-4, -2)$, this is the solution to the system.

12.   E

First, change each equation to $y = mx + b$ form. Since the first equation is already in this form, add $3x$ to each side of the second equation to get $y = 3x + 5$. Each equation has a slope of 3 and different *y*-intercepts. Therefore, the lines are parallel and there is no solution to this system of equations.

13. C

To find the equation of the line that would not intersect with the given line, find the equation of the line that has the same slope. Since the equation $2y + 4 = x$ is equal to $y = \frac{1}{2}x - 2$ in slope-intercept form, the slope of this line is $\frac{1}{2}$ Since choice C also has a slope of $\frac{1}{2}$, it is parallel to the given line and will not intersect it.

14. E

When the first line is written in slope-intercept form, it is equivalent to $y = x + 3$. When the second line is written in slope-intercept form, it is also equal to $y = x + 3$. Since the equations are the same, the lines are coincident and represent the same line when graphed.

15. A

The initial cost is $5.50, so this is the *y*-intercept, or *b*, of the equation. Parking costs $4.00 per additional hour so this number needs to be multiplied by the number of additional hours parked (*x*). This makes 4 the slope, or *m*. Thus, the equation becomes $y = 4x + 5.50$.

16. C

The cost for 8 hours can be calculated by using the formula $y = 4x + 5.50$. Keep in mind that the number of additional hours, or *x*, is equal to 7; the first hour is paid for by the initial cost of $5.50. $y = 4(7) + 5.50 = 28 + 5.50 = \$33.50$.

17. B

The image of the figure in choice B appears to be "flipped" over the diagonal line $y = x$. In addition, each of the coordinates is the inverse of the original figure; (x, y) became (y, x) which indicates a reflection over this line. Choice A is a reflection over the *y*-axis, choice C is a reflection over the *x*-axis, choice D is a rotation of 180°, and choice E is a translation of (4, 2).

18. C

Since the x-value of 3 increases to 6 in the image, this is an increase of 3 units. Since the y-value of -5 increases to -2 in the image, this is also an increase of 3. Therefore, this is a translation of $T_{3, 3}$.

19. A

A rotation of 180° about the origin changes the point (x, y) to $(-x, -y)$. Therefore, the point (5, -6) would become (-5, 6) after this type of rotation.

20. B

Because the difference in the equations is the -1 at the end of the second equation, the image will be a translation of the equation $y = -x^2$ down one unit. The answer choice that shows the same graph just moved down one unit is choice B.

CHAPTER 11

Algebra

Start your study of algebra with this 10-question Building Block Quiz. Remember, the first three questions will revisit some of the important concepts of coordinate geometry. Use the answer explanations following the quiz to help guide you through your study of these topics.

BUILDING BLOCK QUIZ

1. What is the slope of the line containing the points (7, 3) and (5, 2)?

   (A) $\frac{1}{2}$ (B) $\frac{1}{2}$ (C) 1   (D) –2   (E) 2

2. In the linear equation $y = \frac{3}{4}x - 5$, what is the slope?

   (A) 5   (B) –5   (C) $\frac{3}{4}$ (D) $-\frac{3}{4}$ (E) $\frac{4}{3}$

3. The point $(-7, -4)$ is located in which quadrant?

   (A) I   (B) II   (C) III   (D) IV   (E) VI

4. Which of the following best represents the statement "6 more than 3 times the value of x"?

   (A) $6x + 3$ (B) $3(x + 6)$ (C) $3x + 6$
   (D) $6(x + 3)$ (E) $x(6 + 3)$

5. If $\$x\$$ is defined for any integer by the equation $\$x\$ = x(x - 1)$, then what is the value of $\$4\$$?

   (A) 7   (B) 12   (C) 16   (D) 24   (E) 64

6. Which of the following is not an example of a monomial?

   (A) $3x$ (B) -6 (C) $6x^2y$ (D) $x+1$ (E) $9abc$

7. Simplify the following expression: $4(x+3)-(x-5)$.

   (A) $x+8$ (B) $3x+7$ (C) $3x+8$

   (D) $3x+17$ (E) $5x+7$

8. Factor the following expression completely: $3x^2-75$.

   (A) $3(x-5)(x+5)$ (B) $(3x-5)(x+5)$

   (C) $(x-5)(3x+5)$ (D) $3(x-5)^2$

   (E) $3(x-25)$

9. Perform the indicated operation: $\dfrac{3}{x^2}+\dfrac{4a}{x}$.

   (A) $12ax$ (B) $\dfrac{12a}{x}$ (C) $7ax^2$

   (D) $\dfrac{7a}{x}$ (E) $\dfrac{3+4ax}{x^2}$

10. Simplify the expression: $\dfrac{\dfrac{1}{b^2}-1}{b-\dfrac{1}{b}}$.

   (A) b^2-1 (B) $\dfrac{1-b}{b}$ (C) $-b$ (D) $-\dfrac{1}{b}$ (E) $\dfrac{1-b^2}{b-1}$

Answers and Explanations

1. **B**

The slope of the line can be calculated using the formula
$slope = \dfrac{\text{change in } y}{\text{change in } x} = \dfrac{y_1-y_2}{x_1-x_2} = \dfrac{3-2}{7-5} = \dfrac{1}{2}$. The slope of the line is $\dfrac{1}{2}$.

2. **C**

A linear equation in the form $y = mx + b$ has a slope of m and a y-intercept of b. Since the coefficient of x in the equation $y = \dfrac{3}{4}x - 5$ is $\dfrac{3}{4}$, this is the value of m. The slope of the line is $\dfrac{3}{4}$.

3. C

To get to the point $(-7, -4)$, start at the origin and move 7 units to the left and 4 units down. The point lies in quadrant III.

4. C

This question assesses your knowledge of translating statements into algebraic expressions. Break the statement down into smaller parts. The statement "6 more than" tells you to add 6 within the expression. "3 times the value of x" is written as $3x$. The entire expression then becomes $3x + 6$.

5. B

This problem tests your skill in evaluating binary operations. Take the operation $\$x\$ = x(x - 1)$ and substitute 4 for x. $\$4\$ = 4(4 - 1)$ which is equal to $4(3) = 12$.

6. D

This question tests your knowledge of polynomials. A monomial is an algebraic expression that contains only one term. In choice D, there are two terms separated by an addition sign.

7. D

This question assesses your ability to simplify polynomials . To simplify the expression, first multiply using the distributive property to eliminate the parentheses. $4(x + 3) - (x - 5) = 4x + 12 - x + 5$. Then, combine like terms to get the simplified expression. $(4x - x) + (12 + 5) = 3x + 17$.

8. A

Question 8 evaluates your skill in factoring algebraic expressions. First, factor out the greatest common factor (GCF) of 3: $3x^2 - 75 = 3(x^2 - 25)$. Because the binomial within the parentheses is the difference between two perfect squares, factor this difference to $(x - 5)(x + 5)$. The expression, when factored completely, is equal to $3(x - 5)(x + 5)$.

9. E

This question tests your ability to combine rational expressions.

First, get a common denominator of x^2. The expression becomes

$\dfrac{3}{x^2} + \dfrac{4a \times x}{x \times x} = \dfrac{3}{x^2} + \dfrac{4ax}{x^2}$. Add the numerators and keep the common denominator to get $\dfrac{3 + 4ax}{x^2}$.

10. D

This question assesses your proficiency in simplifying complex fractions. First, multiply each term of the expression by the least common denominator of b^2. The expression becomes

$\dfrac{b^2 \times \dfrac{1}{b^2} - 1 \times b^2}{b^2 \times b - \dfrac{1}{b} \times b^2} = \dfrac{1 - b^2}{b^3 - b}$. Now, factor the denominator by taking

out the common factor of b. $\dfrac{1 - b^2}{b(b^2 - 1)}$. When the factors of

$1 - b^2$ and $b^2 - 1$ are cancelled out, -1 is the result. The expression

then becomes $\dfrac{-1}{b}$ or $-\dfrac{1}{b}$.

ALGEBRA

The basic concepts of algebra lay a foundation for all upper-level mathematics, and provide the keys to understanding the world around us. In this chapter, you will embark on a study of the concepts of algebra, including such topics as translating statements into expressions and equations, defined operations, polynomials and factoring, rational expressions, and simplifying complex fractions.

Translating Algebraic Expressions

Algebra is a mathematical language that uses numbers and symbols to create statements and solve problems. The variables, or unknown quantities, that you are solving for are commonly represented by letters such as x or n. Before starting your study of algebraic concepts, it is important to be able to translate word sentences and word problems into algebraic expressions and equations. When translating word phrases into symbols, look for key words that represent certain operations and symbols. On the following page is a chart containing the basic operations and some of the common key words and phrases used with each one.

Add (+)	Subtract (−)	Multiply (×)	Divide (÷)	Equal (=)
Sum, increased by, more than, plus, exceeds	difference, decreased by, less than, minus, reduced	product, multiplied by, of, times	quotient, divided by, into	is, result, total, equal to

It is important to understand the difference between an algebraic expression and an equation.

> **REMEMBER THIS!**
>
> An **algebraic expression** contains numbers, variables, and operations to state a relationship. An **equation** is two algebraic expressions set equal to each other. Therefore, an equation contains an equal sign, where an expression does not.

To convert a statement into an expression or equation, identify the key words for the operations mentioned above and translate the statement. This process is similar to translating from one language to another. For example, to translate the statement "*6 times a number n, added to 4*," you would look for the key word *times* which represents multiplication and the phrase *added to* which indicates addition. Note that *n* represents the unknown number in the statement. Using the numbers in the statement, translate in order from left to right. Therefore, the expression would read $6 \times n + 4$, or $6n + 4$.

Let's try another example. Translate the phrase "*twice the difference of a number n and 4 is 20*." The word *twice* tells you to multiply the remaining part of the statement by 2. Since *difference* is a key word for subtraction, the difference of a number and 4 is written as $n - 4$. The end of the statement "*is 20*" indicates that the expression should be set equal to 20. Therefore, the equation becomes $2(n - 4) = 20$. In this case, the parentheses are important in indicating the correct order of operations.

Most cases of translating expressions are literal; the order of the words/numbers/variables in the sentence is the same as the order in the algebraic expression. However, the exceptions to this rule are the key words *"more than"* and *"less than."* In each case, these phrases change the order of the elements in the statement. For instance, the statement "6 less than x" would be translated as $x - 6$, not $6 - x$. Keep this in mind as you continue your study of translating expressions.

Defined Operations

A **defined operation** is an operation in which symbols (sometimes a strange symbols such as @, or #), are used to represent algebraic expressions. Defined operations often use a combination of common operations like addition, subtraction, multiplication, and division. For example, say you are told that the operation "a & b," where a and b represent integers, is defined as $ab + 4a$. Therefore, to evaluate 2 & 3, substitute 2 for a and 3 for b. The operation becomes:

$$2 \& 3 = 2 \times 3 + 4 \times 2 = 6 + 8 = 14$$

Don't forget to use the correct order of operations when evaluating defined expressions.

FLASHBACK

Refer to chapter 1 to review the correct order of operations.

As another example of a defined operation, say the operation # a # is defined as $a^2 - 1$. So in order to evaluate # 4 #, substitute 4 for a. # 4 # $= 4^2 - 1 = 16 - 1 = 15$.

Practice 1

Match the phrase in the first column with the correct translation in the second column. Answers and explanations are located at the end of the chapter.

1. Three more than twice a number   A. $2n - 3$

2. Three times the sum of a number and 2   B. $2n + 3$

3. Two less than three times a number is 5 C. $3(n + 2)$

4. The product of two and a number, decreased by 3 D. $3n - 2 = 5$

Fill in the blank with the correct solution.

5. If $a \# b$ is defined as $2a^2 - b$, then the value of $1 \# 2$ is

   _____.

6. If $\&c\&$ is defined as $(3c)^2$, then the value of $\&4\&$ is

   _____.

Monomials and Polynomials

When learning the basic elements of algebra, it is important to know the different parts of algebraic expressions.

MATH SPEAK

Numbers, in algebra, are called **constants**.

Letters used to represent unknown quantities are **variables**.

A number that appears in front of a variable (that is connected by multiplication) is called a **coefficient**.

Algebraic expressions containing terms created by the product of constants and variables are called **polynomials**, in which the prefix *poly-* means *many*. A single term is called a **monomial**, in which the prefix of *mono-* means *one*. Some examples of monomials are 3, 4x, –6ab, and $24x^2y$.

Algebraic terms are separated by addition and subtraction. Expressions such as $a + b$ or $3x - 4yz$ are called **binomials** because they contain two terms that do not have exactly the same variables and exponents. Binomials are polynomials with two terms; the prefix *bi-* means *two*.

As you might have guessed, a **trinomial** is a polynomial with three distinct terms, such as $x^2 + 3x - 4$. The prefix *tri-* means *three*.

Adding and Subtracting Polynomials

There are special rules for performing the basic operations with polynomials. The first, which involves adding and subtracting polynomials, is called *combining like terms*.

> **MATH SPEAK**
>
> The phrase **like terms** refers to polynomial terms that contain exactly the same variable(s) and exponent(s).

Some examples of like terms are $3x$ and $4x$, $5ab$ and $-10ab$, and $8r^2t$ and $-9r^2t$. Notice that in like terms, the variables and exponents must be the same, but the coefficients do not have to be the same. When adding and subtracting polynomials, only like terms can be combined. Take the following examples:

$6c + 3c = 9c$

In this example, the variables are the same so the terms can be combined. To combine the terms, add the coefficients of 6 and 3 and keep the variable the same.

$14x^2 - 5x^2 = 9x^2$

This example also contains terms that have the same variables with the same exponents. To simplify the expression, subtract the coefficients of 14 and 5 to get 9. The simplified answer is $9x^2$.

$7x + 8y$

Since the variables here are not the same, these terms cannot be combined. Therefore, the simplified answer is still $7x + 8y$.

Problems that contain grouping symbols sometimes take an additional step to simplify. You must eliminate the grouping symbols before any like terms can be combined.

Take, for example, the expression $(9x - 3y) - (5x - 8y)$. The first step is to use distributive property when eliminating the parentheses. The expression then becomes $9x - 3y - 5x + 8y$. Notice that because $8y$ is a negative term, it becomes positive when the subtraction is distributed. Now, combine like terms to get $9x - 5x - 3y + 8y = 4x + 5y$. This expression cannot be simplified any further since the variable in each term is different.

Practice 2

Using each term no more than once, fill in each blank with the choice from the word bank below that best completes the statement. Answers and explanations are located at the end of the chapter.

binomial   constant   trinomial   monomial   like

coefficient   variable   unlike   distributive property

7. The expression $3x^2 + 5x$ can be described as a _____.

8. In the expression $5x + 6$, the 6 is the _____ term.

9. The polynomial $4c^3 + 8c^3$ contains _____ terms.

10. After the like terms are combined, the polynomial
    $7x - 4x + 8x$ becomes a _____.

11. In the expression $25p^5q$, p represents a _____.

12. In the expression $10b^4$, the number 10 is the _____.

Multiplying Polynomials

When multiplying polynomials, coefficients are multiplied separately from variables. It is important to note the rules for multiplying the exponents of like bases.

> **FLASHBACK**
>
> To review the **laws of exponents**, refer to chapter 7, Powers and Roots.

Here are four situations that occur when multiplying polynomials.

1) When multiplying a *monomial by a monomial*, multiply the coefficients and then follow the rules for multiplying the exponents of like bases. In the example $3x^2 \times 4x^3$, multiply the coefficients and add the exponents of the like bases. Thus, the result is $3x^2 \times 4x^3 = 12x^5$.

2) To multiply a *monomial by a polynomial*, use the distributive property.

> **FLASHBACK**
>
> Refer to chapter 1, Properties of Numbers, to review the distributive property.

For the expression $6y^3(2y + 3)$, multiply the factor of $6y^3$ by both terms inside the parentheses. The expression then becomes $6y^3 \times 2y + 6y^3 \times 3$. Now, simplify each term by multiplying the coefficients and adding the exponents of the like bases. The result is $12y^4 + 18y^3$.

3) To multiply a *binomial by a binomial*, use a special form of the distributive property known by the acronym FOIL.

REMEMBER THIS!

The acronym **FOIL** can be used when multiplying two binomials together. In the expression $(x + a)(x + b)$:

First: Multiply the first term in each binomial $x \times x = x^2$

Outer: Multiply the outer terms in each binomial $b \times x = bx$

Inner: Multiply the inner terms in each binomial $a \times x = ax$

Last: Multiply the last term in each binomial $a \times b = ab$

Final result is $x^2 + bx + ax + ab = x^2 + (a + b)x + ab$

F   O   I   L

Let's practice this process. Using FOIL, simplify $(x + 5)(x - 4)$.

To begin, multiply the first terms $x \times x = x^2$. Then, multiply the outer terms to get $x \times -4 = -4x$. Multiply the inner terms to get $5 \times x = 5x$. Then, multiply the last terms to get $5 \times -4 = -20$. The expression becomes $x^2 - 4x + 5x - 20$. Combine like terms for a final answer of $x^2 + x - 20$.

Try another example with the product $(2x - 3)(3x - 1)$.

The product of the first terms is $6x^2$; the product of the outer terms is $-2x$; the product of the inner terms is $-9x$; the product of the last terms is 3. Combine each of these products to get $6x^2 - 2x - 9x + 3$. Combine like terms to get a final result of $6x^2 - 11x + 3$.

4) To multiply *any polynomial by another polynomial*, you can use the distributive property. In the example $(3x^2 + 4x - 1)(2x^2 - 3x + 3)$, there are two trinomials. Multiply each term from the first trinomial by each term in the second trinomial. The expression becomes:

$3x^2 \times 2x^2 - 3x^2 \times 3x + 3x^2 \times 3 + 4x \times 2x^2 - 4x \times 3x + 4x \times 3 - 1 \times 2x^2 + 1 \times 3x - 1 \times 3$. Perform all of the multiplication first, according to order of operations. The expression becomes:
$6x^4 - 9x^3 + 9x^2 + 8x^3 - 12x^2 + 12x - 2x^2 + 3x - 3$.

Combine like terms to get $6x^4 - x^3 - 5x^2 + 15x - 3$.

When simplifying, some expressions contain a combination of operations. Take the expression $4(x - 5) + 3(2x + 1)$. The first step in solving an expression like this is to multiply using the distributive property to eliminate the parentheses. The expression becomes $4x - 20 + 6x + 3$. The next step is to combine like terms to get a result of $10x - 17$.

Practice 3

Simplify each of the following expressions. Answers and explanations are located at the end of the chapter.

13. $7(x - 3) =$ _____

14. $3x(x^2 + 2x - 4) =$ _____

15. $(x - 3)(x + 4) =$ _____

16. $(x - 2)(3x + 2) =$ _____

17. $(3x - 2) + 5(x + 1) =$ _____

18. $(2x - 1)(3x^2 - 5x + 2) =$ _____

Factoring

In order to simplify algebraic expressions, it is necessary to know how to factor them. The factors of a number or expression are the pieces of the number or expression that multiply together to form the expression. When you are finding the factors of an expression, you are breaking it down into smaller parts.

> ### REMEMBER THIS!
> Finding the **factors** of a number or expression is the opposite of finding its product.

For example, to find factors of the number 6, think of all of the numbers that can be a part of the product of 6. In other words, think of any number that would divide evenly into 6 without a remainder. The factors of 6 are 1, 2, 3, and 6.

To find the factors of an algebraic expression such as $2x^2y$, break down the expression into smaller pieces. $2x^2y$ can be expressed as the product of $2 \times x \times x \times y$. These are all factors of $2x^2y$.

Listed below are various types of expressions and the techniques used to factor these expressions.

Using a Greatest Common Factor

When an expression contains more than one term, it is often helpful to find the GCF factor between the terms in order to simplify. To find the GCF factor of any polynomial, look for common factors in the coefficients, and common variables between each term. For example, in the trinomial $3x^3 - 6x^2 + 9x$, each term has a coefficient factor of 3 and variable factor of x. Factor out $3x$ from each term: $3x^3 - 6x^2 + 9x = 3x(x^2 - 2x + 3)$

The Difference Between Two Squares

There are some special cases of factoring polynomials. One of these cases is finding the difference between two perfect squares, like $x^2 - 25$, $y^2 - 49$, or $4c^2 - 9$. In order to factor the difference between two perfect squares, take the square root of each term. For example, in the binomial $x^2 - a^2$, x is the square root of x^2 and a is the square root of a^2. Then, write the factors in the form $(x - a)(x + a)$. Following this model, the factors of the expressions mentioned above are as follows:

$$x^2 - 25 = (x - 5)(x + 5)$$

$$y^2 - 49 = (y - 7)(y + 7)$$

$$4c^2 - 9 = (2c - 3)(2c + 3)$$

REMEMBER THIS!

When factoring the **difference between two perfect squares**, the factors are in the form $x^2 - a^2 = (x - a)(x + a)$.

$ax^2 + bx + c$

When factoring certain trinomials, the result will often be two binomials. To factor a trinomial in the form $ax^2 + bx + c$, where a, b, and c represent real numbers, take a look at the values of a, b and c. If the value of a is 1, then the second terms in each binomial factor must have a sum of b and a product of c.

For example, take the expression $x^2 + 5x + 6$. The value of $a = 1$, $b = 5$ and $c = 6$. Find the factors of this expression:

First, list factors of 6 (c): 1 and 6, −1 and −6, 2 and 3, and −2 and −3. The only pair of factors that has a sum of 5 (b) is 2 and 3. So now, use the 2 and 3 as the second terms in the binomial factors. This makes the factors of the expression equal to $(x + 2)(x + 3)$.

To check to see if these binomials are the correct factors, use FOIL to multiply them together: $(x + 2)(x + 3) = x^2 + 3x + 2x + 6 = x^2 + 5x + 6$, which was the original trinomial.

Practice this procedure again with the trinomial $x^2 - 5x - 14$.

The value of $a = 1$, $b = -5$ and $c = -14$. Again, first list the factors of −14: 1 and −14, −1 and 14, 2 and −7, and −2 and 7. The only pair of factors in this list that has a sum of −5 is 2 and −7. So use the 2 and −7 as the second terms in the binomial factors. This makes the factors of the expression equal to $(x + 2)(x - 7)$.

Not all instances contain a trinomial where the value of a is 1. Take, for example, the expression $3x^2 + 13x + 4$. Since the values are $a = 3$, $b = 13$ and $c = 4$, you will need to do some trial and error with the factors of 4 to find the binomial factors. Once again, first list the factors of 4: 1 and 4, −1 and −4, 2 and 2, and −2 and -2. Since the value of a is no longer 1 and the first terms need to be factors of $3x^2$, you know the binomials will be in the form $(3x + \_)(x + \_)$. Now, you need to experiment with different placements of the pairs of factors of 4. Start with the factors $(3x + 2)(x + 2)$ and use FOIL to multiply them together. The expression becomes $3x^2$ (F) + $6x$ (O) + $2x$ (I) + 4(L). Combine like terms to get $3x^2 + 8x + 4$. Since this was not the original expression, try again.

This time try 1 and 4 for the factors of 4. The factors become $(3x + 1)(x + 4)$. Now, use FOIL to multiply them together. The expression becomes $3x^2$ (F) $+ 12x$ (O) $+ 1x$ (I) $+ 4$(L). Combine like terms to get $3x^2 + 13x + 4$. Since this was the original expression, these are the correct factors.

> ## REMEMBER THIS!
> When factoring any factorable polynomial in the form $ax^2 + bx + c$ where $a = 1$, the constant terms of the factors have a sum of b and a product of c. If the value of $a \neq 1$, use the factors of c with trial and error to find the factors.

One thing to keep in mind is that not all trinomials are factorable. An example of a non-factorable trinomial is $x^2 + 3x - 1$. There are no factors of -1 that will also have a sum of 3.

Perfect Square Trinomials

A perfect square trinomial is another special case that is the product of two equal binomials. These trinomials are factored in the same way as mentioned above, looking for the sum of b and the product of c. An example of a perfect square trinomial is $x^2 + 10x + 25$. Factors of 25 that will have a sum of 10 are 5 and 5. The factors then become $(x + 5)(x + 5)$, or $(x + 5)^2$. Another example of a perfect square trinomial is $n^2 - 6n + 9$. Since the factors of 9 that will have a sum of -6 are -3 and -3, the factors of the trinomial are $(n - 3)(n - 3)$, or $(n - 3)^2$.

> ## REMEMBER THIS!
> A **perfect square trinomial** has two equal binomial factors. They have two forms and
> are factored as follows:
> $$x^2 + 2ax + a^2 = (x + a)^2 \quad \text{and} \quad x^2 - 2ax + a^2 = (x - a)^2.$$

Factoring Completely

In order for polynomials to be factored completely, the expression must be broken down into its smallest possible factors. This means that there may be two or more steps in order to factor the expression.

Take the expression $2x^2 - 128$. First, look for any common factors between the terms, or the GCF. Each term contains a factor of 2, so the expression becomes $2(x^2 - 64)$. Since the binomial within the parentheses is the difference between two squares, you can break it down even more. The factors now become $2(x - 8)(x + 8)$. The expression is now factored completely.

Another example is the trinomial $3x^3 + 12x^2 - 36x$. First, look for a common factor between the three terms. Each contains a factor of $3x$, so the expression becomes $3x(x^2 + 4x - 12)$ when $3x$ is factored out. Now, see if the trinomial inside the parentheses can be factored. Since the factors of -12 that also combine to a sum of $+4$ are -2 and 6, the factors are $3x(x - 2)(x + 6)$. Now the expression is factored completely.

REMEMBER THIS!

Use these steps to make sure that the expression is factored completely:

1. Factor out the GCF, if it exists.

2. Factor the difference between two squares.

3. Factor the trinomial into two binomials (FOIL).

Practice 4

Answer true or false to each of the following. Answers and explanations are located at the end of the chapter.

19. **T** **F**   The greatest common factor of $4x^4$ and $12x^3$ is $4x^2$.

20. **T** **F**   The factored form of $x^2 - 16$ is $x(x - 16)$.

21. **T** **F**   When factored completely $4x^2 - 24x$ is equal to $4x(x - 6)$.

22. **T F** The two binomial factors of $a^2 - 81$ are $(a - 9)$ and $(a - 9)$.

23. **T F** The trinomial $c^2 - 2c - 35$ is equivalent to the factors $(c - 5)$ and $(c + 7)$.

24. **T F** The expression $x^2 + 6x + 9$ is equivalent to $(x + 3)^2$.

25. **T F** When factored completely, the expression $5x^3 + 15x^2 - 10x$ is equivalent to $5x(x - 1)(x - 2)$.

Rational Expressions

Simplifying

A **rational expression** is an expression that may involve constants and/or variables in the form $\frac{a}{b}$, where b cannot equal 0. When a fraction contains any common factors between the numerator and denominator, it can be reduced or simplified. Take the fraction $\frac{4}{6}$. Since each number has a factor of 2, divide each number by 2 to reduce the fraction: $\frac{4 \div 2}{6 \div 2} = \frac{2}{3}$. This is the simplified form of the fraction.

> **FLASHBACK**
>
> For more review and practice simplifying rational numbers, refer to chapter 3, Fractions and Decimals.

This section will concentrate on rational expressions that contain variables as well as constants. In order to simplify them, you will continue to look for common factors. Let's look at a few examples.

Take the rational expression $\frac{3x}{6x^2}$. When examining the numerator and denominator, each has a factor of $3x$. Divide each by this common factor to simplify the expression: $\frac{3x \div 3x}{6x^2 \div 3x} = \frac{1}{2x}$

In the expression $\frac{4x - 16}{2x - 8}$, both the numerator and denominator need to be factored first in order to see how you can simplify.

$\frac{4x - 16}{2x - 8} = \frac{4(x - 4)}{2(x - 4)}$. Each has a common factor of $x - 4$. When this

factor is divided out of the numerator and denominator, or cancelled,

the result is $\frac{4}{2} = \frac{2}{1}$, which is equal to 2.

In the expression $\frac{x^2 - x - 6}{x^2 - 9}$, both numerator and denominator can be

factored into two binomials. The numerator is a factorable trinomial

and the denominator is the difference between two perfect squares:

$\frac{x^2 - x - 6}{x^2 - 9} = \frac{(x - 3)(x + 2)}{(x - 3)(x + 3)}$. Notice that the common factor of $x - 3$

can be cancelled out. The expression becomes $\frac{(x - 3)(x + 2)}{(x - 3)(x + 3)} = \frac{x + 2}{x + 3}$.

It is important to note at this point that you are only allowed to divide

out, or cancel, common factors. Notice in the last example the x's could

not be cancelled out because they are not factors; in other words, they

are connected to other terms by addition, not multiplication.

> ### REMEMBER THIS!
> When simplifying rational expressions, it is important to factor
> first and be sure to only cancel **factors**.

Adding and Subtracting

When adding and subtracting rational expressions, it is necessary to
find a common denominator. To find the least common denominator
(LCD), find the smallest expression that each denominator will divide
into without a remainder. When this denominator is found, multiply
both the numerator and denominator of the rational expressions by the
missing factor needed to make the LCD. Then combine the
expressions and keep the common denominator.

For example, in order to add the expressions $\frac{2}{x^2} + \frac{5}{x}$, find the least

common denominator of the terms. The LCD of x^2 and x is x^2.

Now, multiply each term by the necessary factor to make each denominator equal to x^2. Since the first fraction already has this denominator, keep it as is. In the second fraction, the denominator of x needs to be multiplied by another factor of x to make it x^2. Therefore, both the numerator and denominator of the second term need to be multiplied by x. $\frac{2}{x^2} + \frac{5}{x} = \frac{2}{x^2} + \frac{5 \times x}{x \times x} = \frac{2}{x^2} + \frac{5x}{x^2}$. Now that there is a common denominator, add the numerators and keep the common denominator to get $\frac{2 + 5x}{x^2}$. Since there are no common factors between the numerator and denominator, the fraction is simplified.

Here are a few more examples of adding and subtracting rational expressions.

$$\frac{8}{x^2 - x} - \frac{3}{x - 1}$$

Factor each expression first to get $\frac{8}{x(x-1)} - \frac{3}{x-1}$. The LCD is $x(x-1)$. Since the first fraction already has this as a denominator, keep it as is. Multiply the numerator and denominator of the second fraction by x to get the common denominator.

$$\frac{8}{x(x-1)} - \frac{3}{x-1} = \frac{8}{x(x-1)} - \frac{3 \times x}{x(x-1)} = \frac{8 - 3x}{x(x-1)}.$$

Since there are no common factors between the numerator and denominator, the fraction is simplified.

$$\frac{x-4}{x^2 + 3x + 2} + \frac{x+1}{x^2 - x - 2}$$

Factor each expression first to get $\frac{x-4}{(x+2)(x+1)} + \frac{x+1}{(x-2)(x+1)}$. The LCD is $(x+2)(x+1)(x-2)$. Therefore, to create two fractions with this common denominator, multiply the numerator and denominator in the first expression by $(x-2)$ and the numerator and denominator in the second expression by $(x+2)$:

$$\frac{(x-2)(x-4)}{(x-2)(x+2)(x+1)} + \frac{(x+2)(x+1)}{(x-2)(x+1)(x+2)}.$$ Use FOIL when

multiplying in each numerator. The expression becomes

$\frac{(x-2)(x-4)}{(x-2)(x+2)(x+1)} + \frac{(x+2)(x+1)}{(x-2)(x+1)(x+2)}$. Combine like terms

in the numerator $\frac{2x^2 - 3x + 10}{(x-2)(x+2)(x+1)}$.

Practice 5

Repetition is the key to math success. Try these questions to help refine your skills with simplifying and combining rational expressions.

26. When simplified completely, the expression $\frac{9x}{x^3}$ is equal to

    _____.

27. When simplified completely, the expression $\frac{6x-2}{3x^2+5x-2}$ is
    equal to _____.

28. If finding the sum of $\frac{7x}{2x^2} + \frac{4}{x^2}$, the common denominator is

    _____.

29. To convert to a common denominator in the difference

    $\frac{5}{x^2-49} - \frac{6x+1}{x-7}$, the numerator and denominator of the
    second fraction would be multiplied by

    _____.

30. After subtracting and simplifying, the expression $\frac{1}{x+2} - \frac{2}{x+1}$
    is equivalent to _____.

Multiplying and Dividing

When performing multiplication and division with rational expressions, a common denominator is not necessary. For these problems, first factor each fraction in the numerator and denominator when possible, and cancel out any common factors between the numerators and denominators. Then multiply across any remaining factors.

Take the example $\frac{xy^2}{ab} \times \frac{b^2}{xy}$. Each part of the expression is the product of factors, so no factoring is necessary. Now, cancel the common factors of xy and b from both fractions and multiply across the remaining factors. The expression becomes $\frac{xy^2}{ab} \times \frac{b^2}{xy} = \frac{yb}{a}$.

Another example is $\frac{x^2 - 25}{2x - 10} \times \frac{2x^2 + 4x}{x + 5}$. First, factor each part the fractions to get $\frac{(x - 5)(x + 5)}{2(x - 5)} \times \frac{2x(x + 2)}{x + 5}$. Cancel the common factors. $\frac{(x - 5)(x + 5)}{2(x - 5)} \times \frac{2x(x + 2)}{x + 5}$. The expression simplifies to

$x(x + 2)$ or $x^2 + 2x$.

The difference between dividing and multiplying rational expressions boils down to one step: dividing by a fraction is the same as multiplying by its reciprocal. Therefore, when the operation is division, simply take the reciprocal of the fraction being *divided by*, and then multiply as explained above. Take, for example, the quotient $\frac{36 - x^2}{4ab} \div \frac{x - 6}{2a}$. The first step is to take the reciprocal of the second fraction and change the operation to multiplication, so the expression becomes

$\frac{36 - x^2}{4ab} \times \frac{2a}{x - 6}$. Now, factor to get $\frac{(6 - x)(6 + x)}{4ab} \times \frac{2a}{x - 6}$.

Cancel any common factors.

Since two of the factors are $x - 6$ and $6 - x$, a little more work needs to be done before canceling. First change $6 - x$ to $-x + 6$, and factor out a -1. This expression now becomes $-1(x - 6)$. Therefore, when you cancel $x - 6$ and $6 - x$ the result is -1, not $+1$. In addition, when the 2 from the numerator of the second fraction is cancelled with the 4 from the denominator of the first fraction, a 2 will remain where the

4 once was, since $4 \div 2 = 2$. $\frac{\overset{1}{(6 - x)}(6 + x)}{\underset{2}{4}ab} \times \frac{2a}{x - 6}$. The product becomes $\frac{-1(6 + x)}{2b} = \frac{-6 - x}{2b}$.

> **REMEMBER THIS!**
> Any factors in the form $\frac{x-a}{a-x}$ will be equal to -1.

Practice 6

For each question, perform the indicated operation. Answers and explanations are located at the end of the chapter.

31. $\dfrac{3x^2}{4y^3} \times \dfrac{12y^2}{9x^2} =$

32. $\dfrac{r^2 - 81}{2t} \times \dfrac{6t^2}{r + 9} =$

33. $\dfrac{2x - 4}{z} \div \dfrac{4 - x^2}{3z} =$

Simplifying Complex Fractions

A complex fraction is a fraction that contains other fraction(s) in the numerator and/or denominator. One way to simplify complex fractions is to find the least common denominator of each fraction within the complex fraction, and multiply each term by this LCD.

In the example $\dfrac{\frac{1}{x}}{\frac{6}{x^2}}$, the LCD is x^2. By multiplying each term by x^2, the

result is $\dfrac{\frac{1}{x} \times x^2}{\frac{6}{x^2}} = \dfrac{x}{6}$ after the common factors are cancelled out.

For another example, look at the complex fraction $\dfrac{\frac{x}{y} + x}{\frac{1}{y^2} + \frac{1}{y}}$.

The LCD is y^2, so multiply each of the four terms by y^2:

$$\frac{\frac{x}{y} \times y^2 + x \times y^2}{\frac{1}{y^2} \times y^2 + \frac{1}{y} \times y^2}$$. Simplify by canceling common factors to get $\frac{xy + xy^2}{1 + y}$.

Factor the numerator to get $\frac{xy(1 + y)}{1 + y}$, which reduces to just xy.

Practice 7

Answer true or false to each statement below. Answers and explanations are located at the end of the chapter.

34. **T   F**   When simplifying the complex fraction $\dfrac{\frac{1}{x^2}}{\frac{2}{3x}}$, the least common denominator is $3x^2$.

35. **T   F**   When simplified, the complex fraction $\dfrac{\frac{1}{a} + \frac{1}{b}}{ab}$ becomes $a + b$.

SUMMARY

Algebra is a critical branch in the world of mathematics, and is a foundation of many mathematical concepts and procedures. In conclusion, remember these important things about working with algebra and its related topics:

- When translating statements into mathematical expressions and equations, use key words to determine the correct operations and symbols.
- When evaluating defined operations, use substitution and evaluate using the correct order of operations.
- When adding and subtracting polynomials, only combine like terms, or terms with exactly the same variables and exponents.
- Use the distributive property (FOIL) when multiplying polynomials.

- When factoring polynomials, always look for the greatest common factor (GCF) of the terms first, the difference between two squares second, and the two binomial factors of a trinomial last.
- When simplifying or performing operations with rational expressions, factor first and only cancel common factors.
- To simplify complex fractions, multiply each term of the fraction by the least common denominator (LCD).

Practice Answers and Explanations

1. B

Twice a number is represented by $2n$, and *three more than* is represented by $+ 3$ to get $2n + 3$.

2. C

The *sum of a number and 2* is represented by $n + 2$, and *three times this sum* is represented by $3(n + 2)$.

3. D

Two less than three times a number is represented by $3n - 2$. Setting this expression equal to 5 results in the equation $3n - 2 = 5$.

4. A

The product of 2 and a number is represented by $2n$. When this amount is decreased by 3, the related expression is $2n - 3$.

5. 0

Substitute $a = 1$ and $b = 2$ into the defined operation.
$2(1)^2 - 2 = 2(1) - 2 = 2 - 2 = 0$.

6. 144

Substitute $c = 4$ into the defined operation. $(3 \times 4)^2 = 12^2 = 144$.

7. binomial

This expression contains two terms.

8. constant

This term contains only the number 6, a constant or unchanging value.

9. like

Each term in the polynomial contains exactly the same variables and exponents on those variables.

10. monomial

Since all of the terms in the expression are like terms, they can be combined to form a single term, or a monomial.

11. variable

Letters, in algebra, are used to represent unknown quantities or variables.

12. coefficient

A number that appears in front of a variable connected by multiplication is known as a coefficient.

13. $7x - 21$

Multiply using the distributive property. $7(x - 3) = 7x - 21$.

14. $3x^3 + 6x^2 - 12x$

Multiply using the distributive property. Be sure to add the exponents when multiplying like bases. $3x(x^2 + 2x - 4) = 3x^3 + 6x^2 - 12x$.

15. $x^2 + x - 12$

Multiply using FOIL. $(x - 3)(x + 4) = x^2 + 4x - 3x - 12$. Combine like terms to get $x^2 + x - 12$.

16. $3x^2 - 4x - 4$

Multiply using FOIL. $(x - 2)(3x + 2) = 3x^2 + 2x - 6x - 4$. Combine like terms to get $3x^2 - 4x - 4$.

17. $8x + 3$

First, multiply the second term using the distributive property. The expression becomes $3x - 2 + 5x + 5$. Then, combine like terms to get $8x + 3$.

18. $6x^3 - 13x^2 + 9x - 2$

Multiply using the distributive property. $(2x - 1)(3x^2 - 5x + 2) = 6x^3 - 10x^2 + 4x - 3x^2 + 5x - 2$. Combine like terms to get $6x^3 - 13x^2 + 9x - 2$.

19. False

The greatest common factor is $4x^3$.

20. False

This is an example of the difference between two perfect squares. The factored form is $(x - 4)(x + 4)$.

21. True

22. False

This is another example of the difference between two perfect squares. The factors are $(a - 9)$ and $(a + 9)$.

23. False

The two binomial factors of this trinomial are $(c + 5)(c - 7)$.

24. True

This is an example of a perfect square trinomial, since the two factors are equal.

25. False

The factors of this trinomial are $5x$ and $(x^2 + 3x - 2)$. This trinomial cannot be factored any further.

26. $\dfrac{9}{x^2}$

Cancel the common factor of x from the numerator and denominator.
$\dfrac{9\cancel{x}}{x^{\cancel{3}}^{2}} = \dfrac{9}{x^2}$.

27. $\dfrac{2}{x + 2}$

First, factor the numerator and denominator, and then cancel out any common factors: $\dfrac{6x - 2}{3x^2 + 5x - 2} = \dfrac{2(3x - 1)}{(3x - 1)(x + 2)} = \dfrac{2}{x + 2}$.

28. $2x^2$

The least common denominator between $2x^2$ and x^2 is $2x^2$.

29. $x + 7$

Since the factors of the denominator of the first fraction are $(x + 7)$ and $(x - 7)$, and the denominator of the second fraction is only $(x - 7)$, then the least common multiple is $(x + 7)(x - 7)$. The missing factor is $(x + 7)$, so this factor is multiplied to the numerator and denominator of the second fraction in order to convert to a common denominator.

30. $\dfrac{-x - 3}{(x + 2)(x + 1)}$

Multiply the numerator and denominator of the first fraction by $(x + 1)$

and the numerator and denominator of the second fraction by $(x + 2)$ to

convert to a common denominator. $\dfrac{1(x + 1)}{(x + 2)(x + 1)} - \dfrac{2(x + 2)}{(x + 1)(x + 2)}$.

Use the distributive property to simplify the numerators.

$\dfrac{x + 1}{(x + 2)(x + 1)} - \dfrac{2x + 4}{(x + 1)(x + 2)}$. Write the numerators over the common

denominator; be sure to subtract both terms in the second fraction.

$\dfrac{x + 1 - 2x - 4}{(x + 1)(x + 2)}$. Combine like terms in the numerator. $\dfrac{-x - 3}{(x + 2)(x + 1)}$

31. $\dfrac{1}{y}$

Cancel any common factors between the numerators and

denominators. $\dfrac{\cancel{3}x^2}{\cancel{4}y^{\cancel{3}}} \times \dfrac{\cancel{12}y^{\cancel{2}}}{\cancel{9}x^{\cancel{2}}}$. Now, multiply the remaining factors of the

numerators and denominators to get $\dfrac{\cancel{3}}{3y} = \dfrac{1}{y}$.

32. $3t(r - 9)$

First factor, and then cancel any common factors between the

numerators and denominators. $\dfrac{r^2 - 81}{2t} \times \dfrac{6t^2}{r + 9} = \dfrac{(r - 9)\cancel{(r + 9)}}{\cancel{2t}} \times \dfrac{\cancel{6}t^{\cancel{2}}}{\cancel{r + 9}}$.

Multiply the remaining factors to get $\dfrac{3r(r - 9)}{1}$, or $3t(r - 9)$.

33. $\dfrac{-6}{2+x}$

First, factor the expressions. $\dfrac{2x-4}{z} \div \dfrac{4-x^2}{3z} = \dfrac{2(x-2)}{z} \div \dfrac{(2-x)(2+x)}{3z}$.

Now, change the operation to multiplication and the second fraction to

its reciprocal. $\dfrac{2(x-2)}{z} \times \dfrac{3z}{(2-x)(2+x)}$. Cancel any common factors

between the numerators and denominators; recall that factors in the

form $x-a$ and $a-x$ cancel to become -1. $\dfrac{2(\overset{-1}{\cancel{x-2}})}{\cancel{z}} \times \dfrac{3\cancel{z}}{(2\cancel{-x})(2+x)}$.

Multiply the remaining factors of the numerators and denominators.

$\dfrac{2x\,(-1)(3)}{2+x} = \dfrac{-6}{2+x}$

34. **True**

Since the denominators are x^2 and $3x$, the least common multiple is $3x^2$.

35. **False**

Multiply each term by the least common denominator of ab.

$\dfrac{\dfrac{1}{a} \times ab + \dfrac{1}{b} \times ab}{ab \times ab}$. This simplifies to $\dfrac{b+a}{a^2b^2}$, which is in its most

simplified form.

CHAPTER 11 TEST

Try the following questions to test your knowledge and understanding
of basic algebraic concepts: translating, factoring, and rational
expressions.

1. Which of the following is the correct translation of the
   statement "five less than six times a number"?

   (A) $5-6n$ (B) $5+6n$ (C) $6-5n$

   (D) $6n-5$ (E) $5n-6$

2. If the operation $*a*$ is defined as $*a* = 2a + 1$, then what is the value of $*5*$?

(A) 9   (B) 10   (C) 11   (D) 25   (E) 26

3. If the operation $b \% c$ is defined as $b \% c = 5b - c$, then what is the value of $3 \% -2$?

(A) –13   (B) 10   (C) 13   (D) 15   (E) 17

4. The algebraic expression $9x^2 + 3x - 2$ can be best described as which of the following?

(A) monomial   (B) binomial   (C) trinomial

(D) pentanomial   (E) none of these

5. In the expression $4x^3y^6z^5$, which of the following numbers represents a coefficient?

(A) 3   (B) 4   (C) 5   (D) 6   (E) all of the above

6. Perform the operation: $4z^2 \times 8z$.

(A) $12z^2$ (B) $12z^3$ (C) $32z^2$ (D) $32z^3$ (E) $32z$

7. What is the product of $7x^2$ and $2x - 5$?

(A) $70x^3$ (B) $14x^2 - 5$ (C) $14x^2 - 35x$

(D) $14x^3 - 35x^2$ (E) $14x^2 - 35x^3$

8. Peter's rectangular living room has dimensions that can be expressed as $x - 3$ and $x + 4$. What is the area of his living room expressed as a polynomial?

(A) $x^2 - 1$ (B) $x^2 - 12$ (C) $x^2 - x - 12$

(D) $x^2 + x + 12$ (E) $x^2 + x - 12$

9. What is the product of the factors $(x - 3)(x + 3)$?

(A) $x^2 + 9$ (B) $x^2 - 9$ (C) $x^2 + 3x - 9$

(D) $x^2 + 6x - 9$ (E) $x^2 - 6x - 9$

10. Multiply: $(x+2)(x^2-3x+4)$.
   (A) x^3-x^2-2x+8 (B) x^3-x^2+x+8
   (C) x^3+x^2+2x+8 (D) $x^3+5x^2-10x+8$
   (E) $x^3-5x^2+10x+8$

11. What are the factors of the binomial $3x^2+9x$?
   (A) $3x^2$ and $9x$ (B) 3 and x^2+3x
   (C) $3x$ and $x+3$ (D) $3x$ and $x+9$
   (E) $3x^2$ and $x+3$

12. Factor the expression: b^2-121.
   (A) $b(b-121)$ (B) $(b-11)(b-11)$
   (C) $(b+11)(b+11)$ (D) $(b-11)(b+11)$
   (E) $(11-b)(b-11)$

13. Factor the expression: $2x^2+5x+2$.
   (A) $(2x+1)(x+2)$ (B) $(2x+2)(x+1)$
   (C) $(2x-2)(x-1)$ (D) $2(x+5)(x+1)$
   (E) $2(x+2)(x+1)$

14. Factor completely: $3x^3-12x^2-15x$.
   (A) $3(x+1)(x-5)$ (B) $3(x-1)(x-5)$
   (C) $3x(x+1)(x-5)$ (D) $3x(x-1)(x+5)$
   (E) $3x^2(x-1)(x-5)$

15. Find the sum: $\dfrac{3c}{4x}+\dfrac{5c}{3x}$.

   (A) $\dfrac{15c}{7x^2}$ (B) $\dfrac{15c^2}{12x^2}$ (C) $\dfrac{11c}{12x}$ (D) $\dfrac{29c}{12x^2}$ (E) $\dfrac{29c}{12x}$

16. Find the difference: $\dfrac{4}{x-1} - \dfrac{10}{x}$.

   (A) $\dfrac{-6x-10}{x(x-1)}$ (B) $\dfrac{6x+10}{x(x-1)}$ (C) $\dfrac{-6x+10}{x(x-1)}$

   (D) $\dfrac{6x-10}{x(x-1)}$ (E) $\dfrac{-6}{x(x-1)}$

17. Find the quotient: $\dfrac{x^2-9x-20}{4x+8} \div \dfrac{x-4}{2x+4}$.

   (A) $\dfrac{2}{x-5}$ (B) $\dfrac{2(x-5)}{x+4}$ (C) $\dfrac{x-5}{x+2}$

   (D) $\dfrac{x+5}{x+4}$ (E) $\dfrac{x-5}{2}$

18. Simplify the expression: $\dfrac{\dfrac{x}{y}+\dfrac{1}{y^2}}{\dfrac{1}{y^2}+1}$.

   (A) $\dfrac{x+1}{y^2}$ (B) $\dfrac{x+y}{1+y}$ (C) $\dfrac{y^2+1}{y^2}$

   (D) $\dfrac{xy+1}{1+y^2}$ (E) $\dfrac{y+1}{y^2+1}$

Answers and Explanations

1. D

The key words *"five less than"* tell you to subtract 5 *from* a quantity, and *"six times a number"* can be written as $6n$. Therefore, the expression becomes $6n - 5$.

2. C

Substitute $a = 5$ into the defined operation and use the correct order of operations. $*5* = 2(5) + 1 = 10 + 1 = 11$.

3. E

Substitute $b = 3$ and $c = -2$ into the defined operation and evaluate using the correct order of operations. $3 \% -2 = 5(3) - (-2) = 15 + 2 = 17$.

4. C

This polynomial has three terms that cannot be combined. A polynomial with three terms is known as a trinomial, the prefix *tri-* meaning "three."

5. B

Since the number 4 is in front of the variables and is connected to them by multiplication, 4 is the coefficient of the term. The 3, 6, and 5 each represent the exponents of the variables used in the expression.

6. D

To find the product, multiply the coefficients and add the exponents of the like bases. The expression becomes $4 \times 8 \times z^{2+1} = 32\,z^3$.

7. D

To find the product, use the distributive property. The expression becomes $7x^2 \times 2x - 7x^2 \times 5 = 14x^3 - 35x^2$.

8. E

To find the area of the room, multiply the dimensions of the room together. To multiply the two binomials, use the acronym FOIL. Multiply the first terms in each binomial (F), multiply the outer terms (O), multiply the inner terms (I), and multiply the last terms in each (L). The product becomes $x^2 + 4x - 3x - 12$. Combine like terms to get $x^2 + x - 12$.

9. B

To multiply the two binomials, use the acronym FOIL. Multiply the first terms in each binomial (F), multiply the outer terms (O), multiply the inner terms (I), and multiply the last terms in each (L). The product becomes $x^2 + 3x - 3x - 9$. Combine like terms to get $x^2 - 9$.

10. A

To multiply the two polynomials, use the distributive property. Multiply each term from the first set of parentheses by the three terms in the second set. The product becomes $x^3 - 3x^2 + 4x + 2x^2 - 6x + 8$. Combine like terms to get $x^3 - x^2 - 2x + 8$.

11. C

Look for the greatest common factor of each term of the binomial, which is $3x$. If $3x$ is factored out of each term, the remaining binomial is $x + 3$. Therefore, the two factors are $3x$ and $x + 3$.

12. D

This is an example of the difference between two perfect squares. The factors are in the form $(x - a)(x + a)$, where x and a are the square roots of the perfect squares. Since the square root of b^2 is b and the square root of 121 is 11, the factors are $(b - 11)(b + 11)$.

13. A

Since the coefficient of x^2 is 2, the factors of this trinomial will be in the form $(2x + a)(x + b)$ where the product of a and b is also 2. Since the only factors of 2 are 1 and 2, try the factors $(2x + 1)(x + 2)$. Checking these factors using FOIL results in the expression $2x^2 + 5x + 2$, which was the original polynomial.

14. C

To factor completely, first look for any common factors between all three terms. Each has a factor of $3x$, so when this is factored out the expression becomes $3x(x^2 - 4x - 5)$. The trinomial within the parentheses can also be factored to $(x + 1)(x - 5)$ because the sum of 1 and -5 is -4 and the product is -5. Therefore, the factors are $3x(x + 1)(x - 5)$.

15. E

Find your common denominator of $3x$ and $4x$, which is $12x$. Multiply the numerator and denominator of the first fraction by 3 and the numerator and denominator of the second fraction by 4 to convert to the common denominator. $\frac{3 \times 3c}{3 \times 4x} + \frac{4 \times 5c}{4 \times 3x} = \frac{9c}{12x} + \frac{20c}{12x}$. Write the numerators over the common denominator and combine like terms to simplify. $\frac{9c + 20c}{12x} = \frac{29c}{12x}$.

16. C

Multiply the numerator and denominator of the first fraction by x and the numerator and denominator of the second fraction by $(x - 1)$ to convert to a common denominator. $\frac{4x}{x(x - 1)} - \frac{10(x - 1)}{x(x - 1)}$. Use the

distributive property to simplify the numerator of the second fraction.
$\frac{4x}{x(x-1)} - \frac{10x-10}{x(x-1)}$. Write the numerators over the common
denominator; be sure to subtract both terms in the second fraction.
$\frac{4x-10x+10}{x(x-1)}$. Combine like terms in the numerator. $\frac{-6x+10}{x(x-1)}$.

17. E

First, factor the expressions. $\frac{x^2-9x-20}{4x+8} \div \frac{x-4}{2x+4} = \frac{(x-4)(x-5)}{4(x+2)} \div$
$\frac{x-4}{2(x+2)}$. Now, change the operation to multiplication and the second
fraction to its reciprocal. $\frac{(x-4)(x-5)}{4(x+2)} \times \frac{2(x+2)}{x-4}$. Cancel any
common factors between the numerators and denominators.
$\frac{(x-4)(x-5)}{\underset{2}{4}(x+2)} \times \frac{2(x+2)}{x-4}$ Multiply the remaining factors of the
numerators and denominators. $\frac{x-5}{2}$.

18. D

Multiply each term by the least common denominator of y^2.
$$\frac{\frac{x}{y} \times y^2 + \frac{1}{y^2} \times y^2}{\frac{1}{y^2} \times y^2 + 1 \times y^2}$$. This simplifies to $\frac{xy+1}{1+y^2}$, which cannot be

simplified further.

CHAPTER 12

Equations

In the preceding chapter, you reviewed the basic concepts of algebra. Now you will extend your knowledge by solving algebraic equations. To prepare for your study, try the following 10-question Building Block Quiz. Some of the questions will review the material from last chapter. Detailed answer explanations are provided to clarify the new concepts, and to prepare you for the upcoming material.

BUILDING BLOCK QUIZ

1. Which of the following expressions is equivalent to the statement "6 less than the product of 5 and a number n"?

   (A) $6-5n$ (B) $5n-6$ (C) $6n-5$
   (D) $5-6n$ (E) $5(n-6)$

2. Which of the following represents the factors of the expression x^2-144?

   (A) $x(x-144)$ (B) $(x+1)(x-144)$ (C) $(x-12)(x+12)$
   (D) $(x-12)(x-12)$ (E) none of these

3. Perform the operation:

   $$\frac{2}{x+1}+\frac{3}{x}$$

   (A) $\frac{5x+3}{x(x+1)}$ (B) $\frac{5}{x(x+1)}$ (C) $\frac{5}{x^2+1}$
   (D) $\frac{5x+1}{x(x+1)}$ (E) $\frac{5x}{x(x+1)}$

4. Solve for x: $-8x = 68$

(A) 8.5 (B) 60 (C) −8.5 (D) −60 (E) 544

5. Solve for x: $3x - 15 = 21$

(A) 2 (B) −2 (C) 36 (D) 12 (E) −12

6. One third of a number, less 7, is 38. What is the number?

(A) 15 (B) 135 (C) $3\frac{1}{3}$ (D) 93 (E) −93

7. Solve for x: $-6(x + 2) = 45$

(A) $-\dfrac{43}{6}$ (B) $-\dfrac{33}{6}$ (C) −9 (D) −9.5 (E) 9.5

8. Solve for x: $-5x + 80 - 7x = -4$

(A) 7 (B) −42 (C) −7 (D) 42 (E) $-6.\overline{3}$

9. Solve the following system of equations:

$2x + y = 14$

$4x + 3y = 36$

(A) (8,3) (B) (6,4) (C) (1,12) (D) (12,1) (E) (3,8)

10. Solve for x: $x^2 - x = 6$

(A) {3,2} (B) {−3,2} (C) {−3,−2}

(D) {3,−2} (E) no solution

Answers and Explanations

1. **B**

This problem reviews translating words into an algebraic expression. Break the words into parts. "The product of 5 and a number n" means to multiply 5 by n, or simply $5n$. "Six less than the product" indicates that the expression is 6 less than the product, hence $5n - 6$.

2. C

This question tests your knowledge of factoring, specifically special factors. This binomial is the difference of two squares, so your answer will follow the model $(x + a)(x - a)$. The factors are $(x - 12)$ and $(x + 12)$.

3. A

This problem reviews your knowledge of combining rational expressions. To add two fractions, find the common denominator, in this case $x(x+1)$. Multiply the numerator and denominator of each fraction by the factor of the common denomator it is missing. The expression becomes $\dfrac{2 \times x}{(x + 1) \times x} + \dfrac{3(x + 1)}{x \times (x + 1)}$

Apply the distributive property and combine like terms in the numerator, to get

$\dfrac{2x + 3x + 3}{x(x + 1)}$ or $\dfrac{5x + 3}{x(x + 1)}$

4. C

This problem reviews one-step equations. Division is the inverse operation of multiplication, so divide both sides of the equation by -8:

$\dfrac{-8x}{-8} = \dfrac{68}{-8}$. Your final answer is $x = -8.5$.

5. D

This problem tests your knowledge of two-step equations. First, you undo the subtraction, and then you undo the multiplication. Addition is the inverse operation of subtraction, so add 15 to both sides of the equation: $3x - 15 + 15 = 21 + 15$, or $3x = 36$. Division is the inverse operation of multiplication, so divide both sides by 3:

$\dfrac{3x}{3} = \dfrac{36}{3}$, or $x = 12$.

6. B

This question tests your skill at translating words into algebra, as well as solving a two-step equation. Translate the words into an algebraic equation: $\frac{1}{3}x - 7 = 38$. First, undo subtraction by adding 7 to both sides, to get $\frac{1}{3}x = 45$. Next, undo the multiplication by dividing both sides of the equation by $\frac{1}{3}$, the equivalent of multiplying both sides by 3. Your answer is $x = 135$.

7. D

Question 7 reviews the use of the distributive property to solve an equation. Apply the distributive property: $-6x - 12 = 45$. Now, undo subtraction by adding 12 to both sides of the equation: $-6x - 12 + 12 = 45 + 12$, or $-6x = 57$. Finally, division is the inverse operation of multiplication, so divide both sides by -6, to get $x = -9.5$.

8. A

This problem deals with the concept of combining like terms to solve an equation. Combine like terms: $-5x - 7x + 80 = -4$, or $-12x + 80 = -4$. Undo addition by subtracting 80 from both sides of the equation: $-12x + 80 - 80 = -4 - 80$, or $-12x = -84$. Since division is the inverse operation to multiplication, divide both sides by -12, to get $x = 7$.

9. E

This question tests your skill at solving a system of equations. Multiply each term of the first equation by -2 to eliminate the variable x: $-4x + -2y = -28$. Now, combine the equations: $(-4x + 4x) + (-2y + 3y) = -28 + 36$. Combine like terms to get $y = 8$. Use the value of 8 for y in the original first equation: $2x + y = 14$, or $2x + 8 = 14$. To solve for x, subtract 8 from both sides of the equation to get $2x = 6$. Finally, divide both sides of the equation by 2 and $x = 3$. The ordered pair is thus (3,8).

10. D

This problem tests your knowledge of quadratic equation solving. Subtract 6 from both sides to get the equation into the form

$ax^2 + bx + c = 0$. The equation is thus $x^2 - x - 6 = 0$. Factor the trinomial: $(x - 3)(x + 2) = 0$. If either of these factors is 0, the equation will be true. So either $x - 3 = 0$, or $x + 2 = 0$. In the first case, $x = 3$ and the second case is that $x = -2$. These are the two solutions to the given equation.

EQUATIONS

Finish your mathematics review in this final chapter by studying different types of equations. You will notice several flashback sidebars in this chapter, because much of the previous material in this book is used to solve algebraic equations.

MATH SPEAK

An **equation** is a mathematical sentence that states that two expressions are equal.

An equation contains at least one variable and an equal sign. The equal sign divides the equation into a lefthand side and a righthand side. When you solve an equation, you simplify expressions, and then perform other operations until the variable is alone on one side of the equation. This is called *isolating the variable*.

REMEMBER THIS!

The goal when solving an equation is to *isolate the variable*.

An equation is presented as an equivalence. Therefore, when you perform a mathematical operation on an equation (such as +, −, ×, or ÷), whatever you do to one side you must do to the other side so the equation stays in balance. Every action or operation you perform on an equation will change the equation into a simpler form, until the variable is isolated.

One-Step Equations

The simplest type of equation to solve is one that requires only one step in order to isolate the variable. This type of equation looks similar to the following: $x + 5 = 12$, $\frac{1}{3}x = 21$, or $\frac{x}{15} = 3$. In all of these examples, only one mathematical operation is present on the side of the equation that contains the variable. In order to isolate the variable, you perform the inverse operation to both sides of the equation. When you perform this inverse operation, the variable will be isolated in one step, and therefore solved.

For example, in the first equation above, you should subtract 5 from both sides of the equation, since subtraction is the inverse operation of addition. The equation then becomes: $x + 5 - 5 = 12 - 5$, which simplifies to $x = 7$. In the second equation above, you should divide both sides by $\frac{1}{3}$, because the inverse operation of multiplication is division. The equation then becomes $\frac{1}{3}x \div \frac{1}{3} = 21 \div \frac{1}{3}$, or $x = 63$.

FLASHBACK

Fraction division was covered in chapter 3, Fractions and Decimals.

After you solve an equation, you can easily check your answer. Just substitute in the answer value for the variable in the original equation and verify that you arrive at a true statement. For example, in the preceding equation, $x + 5 = 12$, the resultant value for x is 7. Evaluate the original equation, using this value: $7 + 5 = 12$; $12 = 12$, a true statement. Check the second equation in a similar manner to verify that you are correct.

The third equation above is read as "x divided by 15 equals 3." The inverse operation to division is multiplication, so you must multiply both sides of this equation by 15 to isolate x: $\frac{x}{15} \times 15 = 3 \times 15$, which

simplifies to $x = 45$.

Perform a check on this answer: $45 \div 15 = 3$; $3 = 3$, a true statement.

Two-Step Equations

Two-step equations are equations in which two operations are present and therefore two steps are required (two inverse operations) to isolate the variable. When performing the inverse operations, always undo the addition or subtraction operation first, and then undo the multiplication or division. For example, in the equation $5x + 12 = -13$, there are two operations present, multiplication and addition. First, subtract 12 from both sides: $5x + 12 - 12 = -13 - 12$. This simplifies to $5x = -25$. Now, divide both sides by 5, to isolate the variable: $\frac{5x}{5} = -\frac{25}{5}$, or $x = -5$.

> **FLASHBACK**
>
> Integer arithmetic was covered in chapter 2, Integers.

Another two-step equation is $\frac{x}{7} - 8 = 20$. The two operations present are division and subtraction; undo the subtraction first by adding 8 to both sides. This will simplify to $\frac{x}{7} - 8 + 8 = 20 + 8$, or $\frac{x}{7} = 28$. Now, multiply both sides by 7 to isolate x: $x = 28 \times 7$, or $x = 196$.

Again, you can easily perform a check on this result. Substitute in the value of 196 for x in the original equation: $\frac{196}{7} - 8 = 20$, which simplifies to $28 - 8 = 20$, or $20 = 20$, a true statement.

382 Math Source

> ## REMEMBER THIS!
>
> When you solve a two-step equation by using the inverse operations, always undo addition or subtraction before multiplication and division.

Sometimes, a problem will be presented in word format, and you must translate the words into an algebraic equation in order to solve. For example: *When you multiply a number by 9 and then subtract 45, the result is 81. What is the number?* First, translate the words into algebra: $9x - 45 = 81$; then solve this two-step equation. Add 45 to both sides of the equation, since addition is the inverse operation of subtraction. $9x - 45 + 45 = 81 + 45$, or, when simplified, $9x = 126$. Now, divide both sides of the equation by 9: $\frac{9x}{9} = \frac{126}{9}$, or $x = 14$.

> ## FLASHBACK
>
> Translating words into algebra was covered in chapter 11, Algebra.

Remember, repetition is the key to mastery! Try these problems that require one or two steps to solve.

Practice 1

Fill in the blanks. Answers and explanations are located at the end of the chapter.

1. When solving an equation, the goal is to _____ the variable.

2. To solve $x - 7 = 10$, you _____ to both sides of the equation.

3. To solve $3x + 9 = 25$, the first step is to _____ from both sides of the equation.

4. Solve for x: $x + 10 = -8$

5. Solve for w: $-2w - 6 = 24$

6. One-fourth of a number less 3 is 17. What is the number?

Solving Equations

Advanced equations require some steps prior to performing the inverse operations explained above. Generally speaking, these additional steps are: applying the distributive property, combining like terms, and moving the variable to one side of the equation.

Applying the Distributive Property

> **FLASHBACK**
>
> The distributive property was covered in chapter 1, Properties of Numbers and in chapter 11, Algebra.

If you are given an equation with parentheses on either side, the first step is to simplify the expression containing the parentheses. Usually, the distributive property is needed to simplify this type of expression. After applying the distributive property, solve as described for a two-step equation. For example, to solve $-4(x + 5) = -64$, first apply the distributive property to get rid of the parentheses: $(-4 \times x) + (-4 \times 5) = -64$, or $-4x - 20 = -64$. Now, add 20 to both sides to get $-4x - 20 + 20 = -64 + 20$, or $-4x = -44$. Finally, divide both sides of the equation by -4: $\frac{-4x}{-4} = \frac{-44}{-4}$, or $x = 11$.

Check your result by substituting in the value of 11 for x: $-4(11 + 5) = -64$, which simplifies to $-4 \times 16 = -64$, and $-64 = -64$.

Combining Like Terms

> ### FLASHBACK
> The phrase **like terms** was defined in chapter 11, **Algebra**.

If an equation has any like terms on either side, you must combine these like terms before performing any inverse operations. For example, in the equation $6x - 25 - 3x = 26$, first combine the like terms. Applying the commutative property, and then combining the variable terms, simplifies the equation to a two-step equation. $6x - 3x - 25 = 26$, or $3x - 25 = 26$. Now, add 25 to both sides to get $3x - 25 + 25 = 26 + 25$, or $3x = 51$. Finally, divide both sides by 3 to isolate the variable: $\frac{3x}{3} = \frac{51}{3}$, and thus $x = 17$.

Verify your work with a check on the original equation: $(6 \times 17) - 25 - (3 \times 17) = 26$. Use the order of operations to simplify: $102 - 25 - 51 = 26$, or $77 - 51 = 26$, or $26 = 26$.

> ### FLASHBACK
> The order of operations was explained in chapter 1, **Properties of Numbers**.

If an equation has both parentheses and the need to combine like terms, the best approach is to first apply the distributive property, then combine like terms, and finally, solve the resulting one or two step equation. For example, to solve the equation $12x - 5(x - 2) = 66$, first distribute the -5 to the terms in parentheses: $12x + -5x - (-10) = 66$. Then, combine like terms: $7x + 10 = 66$. The equation is now a simplified two-step equation. Subtract 10 from both sides: $7x + 10 - 10 = 66 - 10$, or $7x = 56$. Divide both sides of this simpler equation by 7 to isolate the variable: $\frac{7x}{7} = \frac{56}{7}$, or $x = 8$.

A geometry problem can involve the use of equation solving, as in the next example.

If the perimeter of the rectangle below is 112 cm, what are the dimensions of the rectangle?

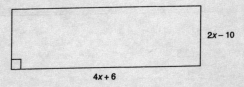

$2x - 10$

$4x + 6$

FLASHBACK

The perimeter of a parallelogram, in this case a rectangle, is covered in chapter 9, Measurement.

The perimeter of a parallelogram is 2 times the base plus 2 times the height, or in this diagram, $2(4x + 6) + 2(2x - 10) = 112$. Apply the distributive property first: $8x + 12 + 4x - 20 = 112$. Now, combine like terms to get $12x - 8 = 112$. Add 8 to both sides: $12x - 8 + 8 = 112 + 8$, or $12x = 120$. Finally, divide both sides by 12: $x = 10$. The value of x is 10 cm, so the base of the rectangle, $4x + 6$, is $(4 \times 10) + 6$ which equals 46. The height is $2x - 10$, or $(2 \times 10) - 10 = 10$. Verify that $(2 \times 46) + (2 \times 10) = 112$. Simplify to get $92 + 20 = 112$, or the true statement $112 = 112$.

Practice 2

Use the number bank to find the solutions to the equations below. Answers and explanations are located at the end of the chapter.

| −12 | −24 | 3.1 | −0.5 | −11 | −3 |
| 4 | 10 | 8 | 18.4 | 36.1 | 4.375 |

7. $-2(x + 8) = 32$

8. $-5(x + 3) = -12.5$

9. $4x - 6 - 7x = 27$

10. $10x - 2(x - 12) = 59$

11. Two times the sum of a number and 4 is 24. What is the number?

12. Given the following triangle, what is the value of the variable x, in degrees?

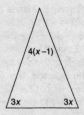

Solving Equations With Variables On Both Sides

After simplifying each side of an equation separately by removing parentheses and combining like terms, the next step is to determine whether the variable appears on both sides of the equation. If so, you must get the variable on one side of the equation. Identify the smaller variable term and add the opposite of this term to both sides of the equation. For example, in the equation $9x + 70 = 3x - 2$ there are no parentheses present and no like terms to combine on the separate sides. There is, however, a variable term on both sides. $3x$ is the smaller of the variable terms, so add the opposite of $3x$ ($-3x$), to both sides of the equation: $9x + -3x + 70 = 3x + -3x - 2$, which simplifies to $6x + 70 = -2$. This is now a two-step equation. Subtract 70 from both sides: $6x + 70 - 70 = -2 - 70$, or $6x = -72$. Finally, divide both sides by 6: $\dfrac{6x}{6} = \dfrac{-72}{6}$, or $x = -12$.

Check this result: $(9 \times -12) + 70 = (3 \times -12) - 2$. This simplifies to $-108 + 70 = -36 - 2$, or $-38 = -38$.

REMEMBER THIS!

When solving an equation, follow these steps to ensure your success:

1) Simplify each side of the equation separately:
   - Apply the distributive property when needed.
   - Combine like terms when needed.

2) Move the variable to one side of the equation.

3) Perform the inverse operations of either addition or subtraction.

4) Perform the inverse operations of multiplication or division.

5) Check your answer, by substituting the value of the variable into the original equation.

When identifying the smaller variable term, remember that a negative number is smaller than a positive number. Take the example of $-11x + 170 = 5x - 22$. In this equation, $-11x$ is the smaller variable term, so add $11x$ to both sides of the equation: $-11x + 11x + 170 = 5x + 11x - 22$. This simplifies to $170 = 16x - 22$, a two-step equation. Now, add 22 to both sides, and then divide both sides by 16: $170 + 22 = 16x - 22 + 22$; $192 = 16x$, and then $\dfrac{192}{16} = \dfrac{16x}{16}$, which simplifies to $x = 12$.

Practice 3

Answer true or false for the following questions.

13. **T** **F** The first step when solving $5(x - 2) = 3x + 10$ is to add 2 to both sides of the equation.

14. **T** **F** The first step when solving $6x - 12 = 5x + 3$ is to add $-5x$ to both sides of the equation.

15. **T** **F** The first step to solve the equation $11x - 3(x + 3) = 7$ is to apply the distributive property.

16. **T** **F** In the equation $x = 3x - 28$, the variable is isolated.

17. **T** **F** In the diagram below, what is the value of the variable x?

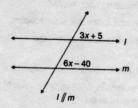

$l \parallel m$

18. **T** **F** Solve for c: $6(c - 4) = 4c + 18$

19. **T** **F** Solve for y: $8y + 12 - 10y = 48$

20. **T** **F** Solve for y: $5y = 3y + 8$

Solving Rational Equations

Rational equations are equations that have fractions in one or more of the terms. One effective way to solve this type of equation is to clear the fractions. Do this by multiplying each term of the equation by the least common denominator of the fractions. For example, to solve the equation $\frac{2x}{5} + \frac{6x}{3} = 21$, multiply each of the terms by the least common multiple of 5 and 3, which is 15: $\times \left(\frac{2x}{5} \cdot \frac{15}{1} \right) + \left(\frac{6x}{3} \times \frac{15}{1} \right) =$

21×15. Cancel common factors to get $(2x \times 3) + (6x \times 5) = 315$, or $6x + 30x = 315$. Combine like terms to get $36x = 315$, and divide both sides by 36. The solution is $x = 315 \div 36$, or $x = 8.75$. Now, check this

answer: $\frac{2 \times 8.75}{5} + \frac{6 \times 8.75}{3} = 21$, or $\frac{17.5}{5} + \frac{52.5}{3} = 21$. Divide

numerators by denominators to get: $3.5 + 17.5 = 21$, or $21 = 21$, a true statement.

FLASHBACK

Simplifying sums and differences of rational numbers was explained in more depth in chapter 11, Algebra.

The least common denominator was covered in chapter 3, Fractions and Decimals.

Practice 4

21. Solve for x: $\frac{2}{3}x + \frac{1}{2} = 8$

22. Solve for x: $\frac{5x}{12} + \frac{5}{6} = 100$

23. Solve for x: $\frac{1}{2}x + \frac{1}{3}x = 1$

Solving Systems of Equations

A system of equations is a set of equations containing two variables, typically x and y. To solve a system of equations, you must find the values of x and y that will make every equation in the system true. There are two algebraic methods of solving a system: the substitution method and the elimination method.

> ### FLASHBACK
>
> Solving systems of linear equations by the graphical method was covered in chapter 10, Coordinate Geometry. In that chapter, it was noted that there can be one solution, no solutions, or an infinite number of solutions to a system of equations. This chapter will focus on the most common situation, when there is one solution to the system.

The Substitution Method

The substitution method of solving a system of equations is to solve one of the equations for one of the variables, say y, and then to substitute in the resultant expression in the second equation for y. The second equation then becomes an equation with one variable, in this case x, which you solve as described earlier in this chapter. For example, to solve the following system of equations:

$$3x + y = -2$$
$$7x - 3y = -26$$

Solve the first equation for y. First, isolate the variable y by subtracting $3x$ from both sides of the equation to get $y = -3x - 2$. Take $-3x - 2$ and substitute it into the second equation in place of y to get $7x - 3(-3x - 2) = -26$. The second equation now contains only one variable, so you can find a solution. Apply the distributive property: $7x + 9x + 6 = -26$. Combine like terms, and subtract 6 from both sides: $16x + 6 - 6 = -26 - 6$. This simplifies to $16x = -32$. Divide both sides by 16 to get $x = -2$. This is the value for x in the system of equations. Now, use one of the equations above and substitute in -2 for x to then solve for y: $3x + y = -2$ becomes $3 \times -2 + y = -2$, or $-6 + y = -2$. Add 6 to both sides to get $y = 4$. The solution to the system is the ordered pair $(-2,4)$. This means that in both equations the solutions are $x = -2$ and $y = 4$.

Keep in mind that there are an infinite number of solutions to each of these equations separately; there is usually one unique solution for the system of two or more equations. Just as for all equations, you can check this solution. It is important to check the solution in both given equations in the system. To check the equation $3x + y = -2$, substitute: $(3 \times -2) + 4 = -2$, or $-6 + 4 = -2$ which is the true statement $-2 = -2$. To check the second equation, use the same procedure: $7x - 3y = -26$ becomes $(7 \times -2) + (-3 \times 4) = -26$. This simplifies to $-14 + -12 = -26$, and then to $-26 = -26$.

The Elimination Method

The elimination method of solving a system of equations involves multiplying each term of one of the equations by some factor, so that when you combine the two equations together, one of the variables will be eliminated. Let's go back and solve the previous system by the elimination method:

$$3x + y = -2$$
$$7x - 3y = -26$$

Look at the variable terms. Since the y term in the first equation has a coefficient of 1, and the y term in the second equation has a coefficient of -3, multiply each of the terms in the first equation by positive 3:

$$9x + 3y = -6$$
$$7x - 3y = -26$$

Now, combine the like terms of the two equations, and notice that the y terms will be eliminated: $(9x + 7x) + (3y + -3y) = -6 + -26$. This simplifies to $16x = -32$, and again, as in the other method above, $x = -2$. From this point, follow the procedures outlined in the substitution method to arrive at the value of 4 for y, and perform the check on both equations.

To solve the next system, look at the equations and find the most convenient variable to eliminate:

$$-4x + 2y = -6$$
$$-2x + 3y = 11$$

It is easy to find a common factor for the coefficients of -4 and -2. To turn the coefficient of x in the the bottom equation to a $+4$, multiply each term of the second equation by -2, so that the x variable terms will be eliminated when the equations are combined:

$$-4x + 2y = -6$$
$$4x + -6y = -22$$

Combine the like terms in the two equations: $(-4x + 4x) + (2y + -6y)$ $= -6 + -22$. This simplifies to $-4y = -28$. Divide both sides by -4, to get $y = 7$. Now, use this value of 7 for y in the first equation to solve for x: $-4x + (2 \times 7) = -6$, or the simplified version $-4x + 14 = -6$. Subtract 14 from both sides, $-4x = -20$, and then divide both sides by -4 to get $x = 5$. The solution to the above system is $(5, 7)$, that is $x = 5$ and $y = 7$. You can check this system in the same manner that you performed the check on the other system of equations.

REMEMBER THIS!

When you check the ordered pair solution to a system of linear equations, be sure to check the solution in every equation in the system.

Some real life situations can be solved using a system of equations. For example, consider the following problem:

Reza bought 6 pounds of fruit, consisting of apples and oranges only. The total cost for the fruit was $6.94. If apples are $0.99 per pound, and oranges are $1.49 per pound, how many pounds of each type of fruit did Reza buy?

Let the variable a represent the number of pounds of apples and the variable b represent the number of pounds of oranges. There is a total of 6 pounds of fruit, so the first equation is $a + b = 6$. The price per pound and total cost is given in the problem. Apples are $0.99 per pound, so your total cost of apples will equal $.99a$; oranges are $1.49 so your total cost of oranges will equal $1.49b$. To find the total cost of both fruits, form the second equation: $0.99a + 1.49b = 6.94$. Using the elimination method, multiply each term of the first equation by -0.99 to eliminate the variable a:

$$-0.99a + -0.99b = -5.94$$
$$0.99a + 1.49b = 6.94$$

Combine the like terms in the two equations to get $(-0.99a + 0.99a)$ $+ (-0.99b + 1.49b) = (-5.94 + 6.94)$. This simplifies to $0.5b = 1$. Divide both sides of the equation by 0.5 to get $b = 2$. Reza bought 2 pounds of oranges. Substitute in the value of 2 for b in the first equation: $a + 2 = 6$. Subtract 2 from both sides to get a, the number of pounds of apples, which is 4 pounds of apples.

Practice 5

Repetition is the key to mastering solving equations! Try either of these two methods to solve the following systems of equations. Answers and explanations are located at the end of the chapter.

24. $6x + y = 16$
    $-4x - 2y = -16$

25. $x + 3y = -1$
    $3x + 2y = 11$

26. The booster club bought jerseys and wind pants for the basketball team. A total of 21 items were bought, at a total cost of $214.50. If the price of one jersey is $8.50, and the price of one pair of pants is $12.50, how many of each item were purchased?

Solving Quadratic Equations

Factoring

> **MATH SPEAK**
>
> A **quadratic equation** is an equation that has the form $ax^2 + bx + c = 0$, where a is non-zero.

To solve a quadratic equation, get all terms of the equation to one side of the equation; the other side should equal zero. Then, try to factor the quadratic into two binomials.

For example, to solve the quadratic equation $x^2 + 12 = 7x$, first add $-7x$ to both sides of the equation to get all terms on the left side. The equation is now $x^2 - 7x + 12 = 0$. Try to factor this trinomial. The factors are $(x - 3)(x - 4)$. Now write the equation using these factors: $(x - 3)(x - 4) = 0$. This equation says the product of two factors is equal to 0. This means that if either the first factor $(x - 3)$ or the second factor $(x - 4)$ is zero, the equation will be true. Solve each of these simple one-step equations to find the two solutions to the quadratic equation. When $x - 3 = 0$, $x = 3$; and when $x - 4 = 0$, $x = 4$. The two solutions to the quadratic are $x = \{3, 4\}$. These solutions can be checked by substituting in the values for x to ensure that the simplified statement is true. Use the original equation of $x^2 + 12 = 7x$. Check 3 as the value of x: $3^2 + 12 = 7 \times 3$, or $9 + 12 = 21$, and $21 = 21$. Check 4 as the value of x: $4^2 + 12 = 7 \times 4$, or $16 + 12 = 28$, and $28 = 28$.

Sometimes a quadratic equation will have two identical solutions. Consider the following quadratic equation: $x^2 + 16x = -64$. Again, the first step is to get all of the terms on one side of the equation. Add 64 to both sides of the equation to get $x^2 + 16x + 64 = 0$. Factor the trinomial. This trinomial is a perfect square, or $(x + 8)(x + 8) = 0$. If either of these factors is 0, the equation will be true. In both cases, the one-step equation is $x + 8 = 0$. When you subtract 8 from both sides, the value of $x = -8$. Now, check this solution in the original equation: $x^2 + 16x = -64$. $(-8)^2 + (16 \times -8) = -64$. This simplifies to $64 + -128 = -64$, which results in the true statement $-64 = -64$.

Sometimes, when solving quadratics with real world problems, you may have to reject one of your two solutions

If the area of the rectangle below is 12 m², what are the possible values of the base and height?

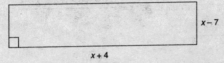

$x - 7$

$x + 4$

Area is base times height, so $(x + 4)(x - 7) = 12$. Use the distributive property (FOIL) to multiply the factors on the left-hand side: $x^2 - 3x - 28 = 12$. Subtract 12 from both sides to get $x^2 - 3x - 28 - 12 = 0$, or $x^2 - 3x - 40 = 0$. Factor the trinomial to get $(x - 8)(x + 5) = 0$. Solve each of these one-step equations to get that either $x = 8$, or $x = -5$. Now, substitute in these values for the base and height to find the possible dimensions. When $x = 8$, the base is $8 + 4$, or 12 m, and the height is $8 - 7$, or 1 m. When $x = -5$, the base would be $-5 + 4$, or -1 m. But there is a problem with this measure—it is not possible to have a negative value for a measurement. Therefore, this solution is rejected for this geometric problem.

The Quadratic Formula

If you cannot factor the resultant trinomial after transforming a quadratic to be equal to 0, then either the quadratic has no real solution or the solutions are not rational numbers. The formula used to solve a quadratic equation is called the **quadratic formula**. This formula will find solutions to any quadratic equation, whether or not the trinomial could have been factored.

> ### REMEMBER THIS!
>
> The **quadratic formula** is a formula used to solve a quadratic equation in the form $ax^2 + bx + c = 0$. The solutions to the equation are found by substituting in the values of the coefficients a, b and c into the formula:
>
> $$x = \frac{-b \pm \sqrt{b^2 - 4ac}}{2a}$$

Notice the plus-minus symbol in the formula. This will yield the two solutions to the quadratic equation. Take as an example the equation $2x^2 + 8x + 3 = 0$. You cannot factor the trinomial, so you must use the quadratic formula, where $a = 2$, $b = 8$ and $c = 3$. Substitute these values into the formula to get: $x = \dfrac{-8 \pm \sqrt{8^2 - 4 \times 2 \times 3}}{2 \times 2}$, or $x = \dfrac{-8 \pm \sqrt{40}}{4}$.

Simplify the radical to get $x = \dfrac{-8 \pm 2\sqrt{10}}{4}$. Divide out a common factor of 2 and the solutions are thus $x = -2 \pm \dfrac{\sqrt{10}}{2}$, or $x = -2 + \dfrac{\sqrt{10}}{2}$, and $x = -2 - \dfrac{\sqrt{10}}{2}$.

FLASHBACK

Simplifying radicals was covered in chapter 7, Powers and Roots.

You can use the quadratic formula on any quadratic equation, even those with a factorable trinomial. But by factoring when possible, you eliminate the need to use this complicated formula.

Sometimes, a quadratic equation has no solutions in the set of real numbers. Consider the following equation: $4x^2 + 2x + 9 = 0$. This trinomial cannot be factored, so use the quadratic formula, where $a = 4$, $b = 2$ and $c = 9$. $x = \dfrac{-2 \pm \sqrt{2^2 - 4 \times 4 \times 9}}{2 \times 4}$, or $x = \dfrac{-2 \pm \sqrt{4 - 144}}{8}$, which simplifies to $x = \dfrac{-2 \pm \sqrt{-140}}{8}$.

In the set of real numbers, the square root of a negative number does not exist. There is no real solution to this quadratic equation.

Practice 6

27. Solve for x: $x^2 + 10x + 25 = 0$

28. Solve for x: $x^2 - 5x = 14$

29. Solve for x: $2x^2 - 7x - 8 = 0$

30. If the area of the rectangle below is equal to 9 square units, what is the value of x?

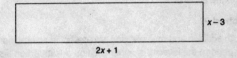

$x - 3$

$2x + 1$

SUMMARY

In conclusion, if you learn only seven things about equations, they should include:

- To solve an equation you must isolate the variable.
- To isolate the variable, first apply the distributive property. Then combine like terms and move the variable to one side. Then undo addition and subtraction, and finally undo multiplication and division.
- You can easily check equation solutions by plugging them into your original equation.
- For rational equations, first multiply all terms by the least common denominator.
- The two algebraic methods for solving a system of equations are the substitution method and the elimination method.
- To solve a quadratic equation, first try to factor the quadratic. If it can be factored, set each of these factors to zero to find the two possible real number answers.
- To solve a quadratic equation that cannot be factored, use the quadratic formula.

Practice Answers and Explanations

1. isolate

Once you have the variable alone on one side of the equation, (isolated) you have solved the equation for the given variable.

2. add 7

Addition is the inverse operation of subtraction, so you add 7 to both sides of the equation to isolate x: $x - 7 + 7 = 10 + 7$, or $x = 17$, the solution to the equation.

3. subtract 9

When both addition and multiplication are present in an equation, you undo, or perform the inverse, of either addition or subtraction first, before you undo multiplication or division. Subtraction is the inverse operation of addition, so the first step is to subtract 9 from both sides of the equation.

4. $x = -18$

Subtract 10 from both sides of the equation.

5. $w = -15$

Add 6 to both sides to get $-2w = 30$. Divide both sides by -2.

6. $x = 80$

The equation is $\frac{1}{4}x - 3 = 17$. Add 3 to both sides of the equation to get $\frac{1}{4}x = 20$. Then, multiply both sides by 4 to isolate the variable.

7. -24

Apply the distributive property to get $-2x - 16 = 32$. Add 16 to both sides to get $-2x = 48$. Divide both sides by -2 to isolate the variable.

8. -0.5

Apply the distributive property to get $-5x - 15 = -12.5$. Add 15 to both sides to get $-5x = 2.5$. Divide both sides by -5 to isolate the variable.

9. -11

Combine the like terms $4x$ and $-7x$ to get $-3x - 6 = 27$. Add 6 to both sides to get $-3x = 33$. Divide both sides by -3 to get $x = -11$.

10. 4.375

Apply the distributive property and then combine like terms: $10x - 2x + 24 = 59$, $8x + 24 = 59$. Subtract 24 from both sides to get $8x = 35$. Divide both sides by 8 to isolate x.

11. 8

The equation is $2(x + 4) = 24$. Apply the distributive property: $2x + 8 = 24$. Subtract 8 from both sides to get $2x = 16$. Divide both sides by 2 to get $x = 8$.

12. 18.4

Recall from chapter 8 that the sum of the degree measure of the angles in a triangle is 180. Use this fact to get the equation $3x + 3x + 4(x - 1) = 180$. Apply the distributive property: $3x + 3x + 4x - 4 = 180$. Combine like terms to get $10x - 4 = 180$. Add 4 to both sides, and divide both sides by 10 to get $x = 18.4$

13. False

The first step is to apply the distributive property.

14. True

15. True

16. False

There is a variable on both sides of this equation.

17. 15

The two angles marked are corresponding angles, described in chapter 8. Their measures are equal, so set up the equation $3x + 5 = 6x - 40$. Subtract $3x$ from both sides to get $5 = 3x - 40$. Add 40 to both sides: $45 = 3x$. Divide both sides by 3 to get $x = 15$.

18. 21

First, apply the distributive property to get $6c - 24 = 4c + 18$. Add $-4c$ to both sides: $2c - 24 = 18$. Add 24 to both sides, to get $2c = 42$. Divide both sides by 2 to isolate c, thus $c = 21$.

19. −18

First, combine like terms to get $-2y + 12 = 48$. Then, subtract 12 from both sides: $-2y = 36$. Finally, divide both sides by −2 to get $y = -18$.

20. 4

Add $-3y$ to both sides of the equation: $5y + -3y = 3y + -3y + 8$, or $2y = 8$. Divide both sides by 2 to isolate y, and $y = 4$.

21. 11.25

Multiply each term by the LCM of 2 and 3, which is 6, to get $4x + 3 = 48$. Subtract 3 from both sides: $4x + 3 - 3 = 48 - 3$, or $4x = 45$. Finally, divide both sides by 4 and x is isolated: $x = 11.25$.

22. 238

Multiply each term by 12: $5x + 10 = 1200$. Subtract 10 from both sides to get $5x = 1190$. Finally, divide both sides by 5: $x = 238$.

23. $\frac{6}{5}$

Multiply all terms in the equation by the least common denominator of 6: $\left(\frac{1}{2}x \times 6\right) + \left(\frac{1}{3}x \times 6\right) = 1 \times 6$, which simplifies to $3x + 2x = 6$.

Combine like terms: $5x = 6$. Divide both sides by 5: $x = \frac{6}{5}$.

24. (2,4)

You can multiply each term of the first equation by 2, and use the elimination method:

$$12x + 2y = 32$$
$$-4x - 2y = -16$$

Combine the two equations: $(12x + -4x) + (2y + -2y) = (32 + -16)$, or $8x = 16$. Divide each side of this simplified equation by 8 to get $x = 2$. Now, substitute the value of 2 into the original first equation: $6 \times 2 + y = 16$, or $12 + y = 16$. Subtract 12 from both sides to get $y = 4$. The ordered pair is (2,4).

25. (5, –2)

Use the substitution method or the elimination method. Using the substitution method, solve the first equation for x, then use this value to find y. The first equation becomes $x = -3y - 1$. Substitute in this value for x in the second equation: $3(-3y - 1) + 2y = 11$. Apply the distributive property: $-9y - 3 + 2y = 11$. Now, combine like terms to get $-7y - 3 = 11$. Add 3 to both sides: $-7y = 14$. Finally, divide both sides by -7 to get $y = -2$. Use the value of -2 in the first equation to find the value of x: $x + 3 \times -2 = -1$, or $x - 6 = -1$. Add 6 to both sides, and $x = 5$. The ordered pair is (5,–2).

26. 12 jerseys and 9 wind pants

Let j represent the number of jerseys, and w represent the number of wind pants. The total number of items bought is 21, so the first equation is $j + w = 21$. The total cost is 214.50, so the second equation is $8.5j + 12.5w = 214.50$. Multiply the first equation by -8.5 to eliminate j:

$-8.5j + -8.5w = -178.50$

$8.5j + 12.5w = 214.50$

Combine the two equations: $(-8.5j + 8.5j) + (-8.5w + 12.5w) = -178.50 + 214.50$. Simplify to get $4w = 36$. Divide both sides by 4 to get $w = 9$. A total of 21 items were bought, so the number of jerseys is $21 - 9 = 12$.

27. {–5}

The lefthand side of this equation is a perfect square trinomial. Factor it to get $(x + 5)(x + 5) = 0$. If either of the two factors are 0, the equation is true. In both cases, this is when $x = -5$.

28. {7,–2}

Subtract 14 from both sides to get the equation into the form of $ax^2 + bx + c = 0$: $x^2 - 5x - 14 = 0$. Now, factor the trinomial on the left into its factors, to get $(x - 7)(x + 2) = 0$. If either of these factors is 0, the equation will be true. Therefore, $x = 7$, and $x = -2$ are the two solutions to the equation.

29. $\left\{ \dfrac{7 + \sqrt{113}}{4}, \dfrac{7 - \sqrt{113}}{4} \right\}$

The trinomial on the left hand side of the equation cannot be factored. Use the quadratic formula to solve for x, with $a = 2$, $b = -7$, and $c = -8$.

$$\dfrac{-(-7) \pm \sqrt{(-7)^2 - (4)(2)(-8)}}{(2)(2)} = \dfrac{7 \pm \sqrt{49 + 64}}{4} =$$

$$\dfrac{7 \pm \sqrt{113}}{4}$$

30. {4}

Area is base times height, and is equal to 9 square units. Set up the equation: $(x - 3)(2x + 1) = 9$. Multiply, using the distributive property (FOIL) : $2x^2 + x - 6x - 3 = 9$. Combine like terms: $2x^2 - 5x - 3 = 9$, and then subtract 9 from both sides of the equation to solve. $2x^2 - 5x - 3 - 9 = 9 - 9$, which simplifies to $2x^2 - 5x - 12 = 0$. The left hand side can be factored: $(2x + 3)(x - 4) = 0$. For the equation to be true, either $2x + 3 = 0$, or $x - 4 = 0$. Subtract 3 from both sides of the first equation to get $2x = -3$. Divide both sides by 2, and $x = -\dfrac{3}{2}$. The second equation will be true when $x = 4$. The quadratic equation has two solutions: $\left\{ -\dfrac{3}{2}, 4 \right\}$. However, when $x = -\dfrac{3}{2}$, the base value of the rectangle is $-\dfrac{3}{2} - 3$, which is a negative number. Since length cannot be a negative value, the solution of $-\dfrac{3}{2}$ is rejected, and the only solution to this geometry application is 4.

CHAPTER 12 TEST

Try the following questions to see what you've learned about equation solving. Following the questions are complete answer explanations to help you assess your understanding. Unless otherwise noted, give each answer to the nearest tenth.

1. Solve for x: $x + 21 = 15$

   (A) 315   (B) 36   (C) −36   (D) 6   (E) −6

2. Solve for x: $-12x = 138$

   (A) 11.5   (B) −11.5   (C) 150   (D) 126   (E) −150

3. Solve for y: $\frac{1}{4}y + 8 = 12$

   (A) 2   (B) 4   (C) 5   (D) 8   (E) 16

4. Solve for x: $3(x - 2) = 36$

   (A) 14   (B) 10   (C) 12.7   (D) 13.7   (E) 10.3

5. Solve for b: $-4(b - 6) = -60$

   (A) −9   (B) 21   (C) −21   (D) 9   (E) 13.5

6. If you multiply 5 times the sum of a number and 13, the result is 100. What is the number?

   (A) 20   (B) 33   (C) 7   (D) 17.4   (E) 22.6

7. Solve for z: $3z + 23 + 6z = 104$

   (A) 14.1   (B) 25   (C) 72   (D) 9   (E) 3.4

8. Solve for x: $4x - 8 = x + 1$

   (A) 1.8   (B) 3   (C) $2.\overline{3}$ (D) 1.4   (E) 2.25

9. $25 - 4(x + 3) = 4x + 33$

   (A) −5   (B) −7.625   (C) −0.75   (D) 2.5   (E) −2.5

10. In the figure below, what is the value of the variable x?

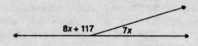

$8x + 117$ $7x$

(A) 117   (B) 4.2   (C) 23.4   (D) –117   (E) –4.2

11. In the figure below, lines l and m are parallel. What is the value of the variable x?

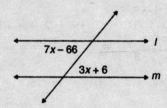

$7x - 66$

$3x + 6$

(A) 6   (B) 7.2   (C) 15   (D) 18   (E) –6

12. Solve for x: $\dfrac{4x}{7} - \dfrac{2}{5} = 6$

(A) $2.\overline{2}$ (B) 0.4   (C) 11.2   (D) 1   (E) 9.8

13. Solve for x: $\dfrac{1}{6}x + \dfrac{1}{8}x = 1$

(A) 3.43   (B) 12   (C) 7   (D) 4.5   (E) 4.27

14. Solve the system of equations:

$3x + y = 4$

$2x - 8y = -58$

(A) (1,–7)   (B) (–7,1)   (C) (7,–1)
(D) (7,1)   (E) (–1,7)

15. Solve the system of equations:

    $y = 6x + 10$

    $2x + 3y = 110$

    (A) (5,40) (B) (7,52) (C) (4,34)

    (D) (34,4) (E) (40,5)

16. One hundred and twenty favors were purchased for the Valentine's Day dance, in either red or pink. The pink favors cost $2.00 each and the red favors cost $1.75 each. If the total cost of the favors was $222.50, how many pink favors were purchased?

    (A) 50 (B) 70 (C) 60 (D) 111 (E) 9

17. Twenty-five adults and children went to a restaurant for a brunch. The cost of the brunch is $11.95 for adults, and $3.95 for children. If the total bill was $266.75, how many adults attended the brunch?

    (A) 4 (B) 46 (C) 21 (D) 11 (E) 22

18. Solve for x: $x^2 - 14x = -49$

    (A) $\{-7\}$ (B) $\{7\}$ (C) $\{-7,7\}$ (D) $\{40,9\}$ (E) $\{-40,9\}$

19. Solve for x: $x^2 - 36 = 0$

    (A) $\{-18\}$ (B) $\{6\}$ (C) $\{-6\}$ (D) $\{18\}$ (E) $\{6,-6\}$

20. Solve for x: $2x^2 - 3x - 5 = 15$

    (A) $\left\{-\dfrac{5}{2}, 4\right\}$ (B) $\left\{-\dfrac{5}{2}, -4\right\}$ (C) $\left\{\dfrac{5}{2}, 4\right\}$

    (D) $\left\{\dfrac{5}{2}, -4\right\}$ (E) $\left\{\dfrac{3 + \sqrt{89}}{4}, \dfrac{3 - \sqrt{89}}{4}\right\}$

21. Solve for x: $3x^2 + 2x - 15 = 0$

(A) $\{5, -3\}$ (B) $\left\{ \dfrac{-1 + \sqrt{46}}{3}, \dfrac{-1 - \sqrt{46}}{3} \right\}$ (C) $\{-5, 3\}$

(D) $\{-15, 1\}$ (E) no real number solution

22. If the area of the triangle below is 8 square units, what is the value of the variable x?

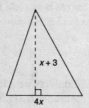

(A) 4 (B) 1 (C) $\{1, -4\}$ (D) $\{2, -1\}$ (E) $\{1, 4\}$

Answers and Explanations

1. E

Since subtraction is the inverse operation to addition, subtract 21 from both sides of the equation: $x + 21 - 21 = 15 - 21$. This simplifies to $x = -6$.

2. B

The inverse operation of multiplication is division; therefore, divide both sides of the equation by -12: $-12x \div -12 = 138 \div -12$, or $x = -11.5$

3. E

In a two-step equation, undo the addition before multiplication. The inverse of addition is subtraction; subtract 8 from both sides: $\frac{1}{4}y + 8 - 8 = 12 - 8$. This simplifies to $\frac{1}{4}y = 4$. Now, undo multiplication by dividing both sides by $\frac{1}{4}$, which is the same as multiplying by 4: $4 \times \frac{1}{4}y = 4 \times 4$. This simplifies to $y = 16$.

4. A

First, apply the distributive property to get $3x - 6 = 36$. Add 6 to both sides of the equation: $3x - 6 + 6 = 36 + 6$, or $3x = 42$. Divide both sides of the equation by 3, to get $x = 14$.

5. B

First, apply the distributive property: $(-4 \times b) - (-4 \times 6) = -60$, $-4b - (-24) = -60$, which simplifies to $-4b + 24 = -60$. Subtract 24 from both sides of the equation, to get $-4b + 24 - 24 = -60 - 24$, or $-4b = -84$. Now, divide both sides by -4 to get $b = 21$.

6. C

Translate the words into an algebraic equation: $5(x + 13) = 100$. Use the distributive property to get $5x + 65 = 100$. Subtract 65 from both sides to get $5x = 35$. Finally, divide both sides by 5 to get $x = 7$.

7. D

Combine like terms to get $3z + 6z + 23 = 104$, or $9z + 23 = 104$. Subtract 23 from both sides of the equation to get $9z + 23 - 23 = 104 - 23$, or $9z = 81$. Divide both sides by 9 to isolate the variable and find that $z = 9$.

8. B

Add $-x$ to both sides to get the variable on one side of the equation: $4x + -x - 8 = x + -x + 1$. Combine like terms to get $3x - 8 = 1$. Add 8 to both sides of the equation: $3x - 8 + 8 = 1 + 8$, or $3x = 9$. Divide both sides by 3 and $x = 3$.

9. E

The first step is to apply the distributive property: $25 - 4x - 12 = 4x + 33$. Combine like terms on the left hand side: $25 - 12 - 4x = 4x + 33$ to get $13 - 4x = 4x + 33$. Add $4x$ to both sides to get the variable on one side of the equation: $13 - 4x + 4x = 4x + 4x + 33$, or $13 = 8x + 33$. Subtract 33 from both sides to get $-20 = 8x$. Now, divide both sides by 8: $x = -20 \div 8$, or $x = -2.5$.

10. B

These angles are a linear pair, supplementary angles, in which the sum of the degree measures is 180°. Set up the equation: $8x + 117 + 7x = 180$. Combine like terms to get $15x + 117 = 180$. Subtract 117 from both sides of the equation: $15x = 63$. Divide both sides of the equation by 15 to isolate x and $x = 4.2$.

11. D

The angles marked are alternate interior angles, so their measures are equal. Set up the equation: $7x - 66 = 3x + 6$. Add $-3x$ to both sides of the equation to get the variable on one side: $7x + -3x - 66 = 3x + -3x + 6$. Combine like terms and add 66 to both sides: $4x = 72$. Divide both sides by 4 and $x = 18$.

12. C

Multiply all terms in this equation by 35, the LCM of 7 and 5. This will clear the fractional terms of the equation. The equation becomes $20x - 14 = 210$. Add 14 to both sides to get $20x = 224$. Divide both sides by 20 and $x = 11.2$.

13. A

Multiply each term by 24, the LCM of 6 and 8. The equation becomes $4x + 3x = 24$. Combine like terms: $7x = 24$. Divide both sides by 7 and x is approximately 3.43.

14. E

Because the coefficient of y in the second equation is 1, multiply the first equation by 8 to eliminate the y variable by combining the two equations:

$$24x + 8y = 32$$
$$2x - 8y = -58$$

$(24x + 2x) + (8y + -8y) = 32 + -58$, when simplified, becomes $26x = -26$. Divide both sides by 26 and $x = -1$. Substitute in the value of -1 for x in the original first equation to find the value of y: $3(-1) + y = 4$, or $-3 + y = 4$. Add 3 to both sides of the equation and $y = 7$. The solution is the ordered pair $(-1, 7)$.

15. C

The first equation is already solved for the variable y, so use the substitution method and substitute in the expression $6x + 10$ for the value of y in the second equation: $2x + 3(6x + 10) = 110$. Apply the distributive property: $2x + 18x + 30 = 110$. Next, combine like terms: $20x + 30 = 110$. Subtract 30 from both sides of the equation to get $20x = 80$. Divide both sides by 20 and $x = 4$. Using the value of 4 for x, use the first equation to find y: $6(4) + 10$, or $y = 34$. The solution is the ordered pair $(4, 34)$.

16. A

Let r represent the number of red favors, and p represent the number of pink favors. There were a total of 120 favors, so the first equation is $r + p = 120$. Using the cost and total purchase price, the second equation is $2.00p + 1.75r = 222.50$. The problem asks for the number of pink favors purchased, so solve the first equation for r in terms of p, which is $r = 120 - p$. Substitute in this expression for r in the second equation: $2.00p + 1.75(120 - p) = 222.50$. Apply the distributive property to get $2p + 210 - 1.75p = 222.50$. Combine like terms: $0.25p + 210 = 222.50$. Subtract 210 from both sides to get $0.25p = 12.5$. Divide both sides by 0.25, and $p = 50$.

17. C

Let a represent the number of adults, and c represent the number of children. There were a total of 25 people, so the first equation is $a + c = 25$. Using the cost and total bill, the second equation is $11.95a + 3.95c = 266.75$. The problem asks for the number of adults, so solve the first equation for c in terms of a, which is $c = 25 - a$. Substitute in this expression for c in the second equation: $11.95a + 3.95(25 - a) = 266.75$. Apply the distributive property to get $11.95a + 98.75 - 3.95a = 266.75$. Combine like terms: $8a + 98.75 = 266.75$. Subtract 98.75 from both sides to get $8a = 168$. Divide both sides by 8, and $a = 21$.

18. B

First, add 49 to both sides to get the equation in the correct form: $x^2 - 14x + 49 = 0$. The lefthand side is a perfect square trinomial when factored: $(x - 7)(x - 7) = 0$. When either of these factors are 0 the equation is true, so the solution is $x = 7$.

19. E

The binomial on the left hand side is, when factored, the difference of two squares: $(x + 6)(x - 6) = 0$. The equation will be true when either of the factors is equal to 0; this is when either $x = 6$, or $x = -6$.

20. A

First, subtract 15 from both sides of the equation to get the equation in the correct form: $2x^2 - 3x - 5 - 15 = 15 - 15$, or $2x^2 - 3x - 20 = 0$. Factor the left hand side of the equation to get $(2x + 5)(x - 4) = 0$. The equation will be true when either $2x + 5 = 0$, or $x - 4 = 0$. Subtract 5 from both sides of the first equation to get $2x = -5$. Divide both sides by 2 and the first value for x is $-\frac{5}{2}$. The second equation, $x - 4 = 0$, will be true when $x = 4$, which is the second value for x.

21. B

This trinomial cannot be factored, so you must use the quadratic formula, where $a = 3$, $b = 2$ and $c = -15$:

$\frac{-(2) \pm \sqrt{2^2 - (4)(3)(-15)}}{2(3)}$, which is then $\frac{-2 \pm \sqrt{4 + 180}}{6} =$

$\frac{-2 \pm \sqrt{184}}{6}$. Simplify the radical: $\frac{-2 \pm 2\sqrt{46}}{6}$. Now, cancel out the common factor of 2 in each term of the numerator and the denominator to get the two values of x: $\frac{-1 \pm \sqrt{46}}{3}$.

22. B

The formula for the area of a triangle is $\frac{1}{2}b \times h$. In this triangle, the base (b), is $4x$ and the height (h) is $x + 3$. The area of the triangle is thus $\frac{1}{2}(4x)(x+3)$, which equals 8 square units. Multiply $\frac{1}{2} \times 4x$ to get $2x$, so therefore $2x(x + 3) = 8$. Distribute this $2x$ to get $2x^2 + 6x = 8$. Now, subtract 8 from both sides: $2x^2 + 6x - 8 = 0$. Factor the left hand side: $(2x - 2)(x + 4) = 0$. The equation will be true if either of the factors equals 0. This will be when either $x = 1$ or $x = -4$. Even though the equation does have two solutions, the value of -4 must be rejected, because it is not possible to have a negative length. The solution is $x = 1$.

CHAPTER 13

Math Source Cumulative Test

Congratulations! You have made your way through the basic building blocks of math, and have built a solid understanding of math concepts ranging from number properties to decimals to coordinate geometry, and more! The following cumulative review tests the skills you have learned throughout *Math Source*. Try the following questions to assess how far you've come. Answers with detailed explanations follow to help further you in your study and understanding.

1. On a series of plays, a football team loses 9 yards on the first play, loses 5 more yards on the second play, gains 12 yards on the third play, and loses 2 yards on the fourth play. If the first play started at the 22nd yard line, what yard line did they end up on after the fourth play?

   (A) 18th   (B) 4th   (C) 36th   (D) 6th   (E) 26th

2. Factor the expression: $x^2 + 8x + 16$.

   (A) $(x+4)(x-4)$ (B) $(x+4)^2$ (C) $(x+1)(x+16)$

   (D) $(x+8)^2$ (E) $(x+2)(x+8)$

3. Find the product: $\dfrac{5x-15}{9y^2} \times \dfrac{18y}{x-3}$.

   (A) $\dfrac{10}{y}$ (B) $10y$ (C) $\dfrac{7}{y}$ (D) $\dfrac{10y}{x-3}$ (E) $\dfrac{7y}{x-3}$

4. Evaluate: $(-90 \div 45) \times -3$

   (A) -6 (B) -135 (C) 15 (D) 6 (E) 135

5. A boat sights the top of a lighthouse, known to be 45 feet tall. The angle of elevation is 30°. How far is the boat from the base of the lighthouse?

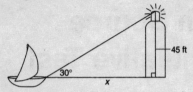

45 ft

30°

x

   (A) 54.1 ft   (B) 90.0 ft   (C) 75.0 ft
   (D) 77.9 ft   (E) 52.0 ft

6. How many square feet of carpeting is needed to cover a floor that is 32 feet by 28 feet?
   (A) 120 ft^2 (B) 896 ft^2 (C) 1,792 ft^2
   (D) 60 ft^2 (E) 1,808 ft^2

7. Which concept is demonstrated in $7 \times (12 - 3) = (7 \times 12) - (7 \times 3)$?
   (A) order of operations   (B) distributive property
   (C) associative property   (D) commutative property
   (E) none of the above

8. What is the slope of the line that contains the points $(9, 2)$ and $(6, -1)$?
   (A) 3   (B) –3   (C) –1   (D) $\frac{1}{3}$ (E) 1

9. What is the solution to the system of equations below?
$$x = 4$$
$$y = x + 5$$

   (A) $(5, 0)$ (B) $(5, 9)$ (C) $(9, 4)$
   (D) $(4, 0)$ (E) $(4, 9)$

10. The formula to convert from degrees Fahrenheit to degrees Celsius is $C = \dfrac{5}{9}(F - 32)$, where F is the degrees in Fahrenheit. What are the degrees Celsius of a temperature of 104 degrees Fahrenheit?

    (A) 40 (B) 54 (C) 360 (D) 3,240 (E) 8

11. Which of the following numbers is divisible by 9?
    (A) 996 (B) 17,145 (C) 980
    (D) all of the above (E) none of the above

12. $3\dfrac{5}{8} - 4\dfrac{1}{4} =$

    (A) $7\dfrac{7}{8}$ (B) $\dfrac{5}{8}$ (C) $\dfrac{1}{2}$ (D) $-\dfrac{5}{8}$ (E) $1\dfrac{3}{8}$

13. Kate had $19.03 left on her gift card. She used the card to purchase an item for $9.99. How much money is now left on the gift card?
    (A) $90.40 (B) $ 9.05 (C) $ 0.90
    (D) $ 0.09 (E) $ 9.04

14. Which of the following sets of side lengths could be the measures of the sides of a triangle?
    (A) { 6, 7, 14 }
    (B) { 5, 5, 10 }
    (C) { 6, 8, 10 }
    (D) All the above sets are possible.
    (E) None of the above sets are possible.

15. If you take one-half of a number and add 19, the result is 44. What is the number?
    (A) 50 (B) 25 (C) 12.5 (D) 31.5 (E) 22

16. Which of the following are true?
    (A) All rectangles are parallelograms.
    (B) All trapezoids are quadrilaterals.
    (C) All squares are rhombuses.
    (D) All of the above are true.
    (E) None of the above are true.

17. Simplify $\dfrac{(2^4 \times 2^5)}{2^3}$.
    (A) 2^6 (B) 2^9 (C) 2^{12} (D) 2^{17} (E) 2^{23}

18. $2\sqrt{150} =$
    (A) $2\sqrt{6}$ (B) $5\sqrt{6}$ (C) $15\sqrt{10}$
    (D) $10\sqrt{10}$ (E) $10\sqrt{6}$

19. Solve for x: $5x^2 - 4x + 2 = 0$
    (A) $\left\{\dfrac{2}{5}\right\}$ (B) $\left\{\dfrac{2}{5}, \dfrac{1}{5}\right\}$ (C) $\{2,1\}$
    (D) $\{-2,-1\}$ (E) no real number solution

20. In a local high school band, the ratio of trombone players to trumpet players is 3 to 4. Which of the following cannot be the total number of musicians that play either the trombone or trumpet?
    (A) 14 (B) 84 (C) 126 (D) 196 (E) 204

21. For a $4,500.00 certificate of deposit, the bank will pay 8.5% simple interest if it is invested for 18 months. How much interest will the certificate accrue?
    (A) $6,885.00 (B) $38,250.00 (C) $573.75
    (D) $6,750.00 (E) $688.50

22. When rolling one die, what is the probability of landing on a prime number?

(A) $\frac{1}{4}$ (B) $\frac{1}{2}$ (C) $\frac{2}{3}$ (D) $\frac{1}{3}$ (E) $\frac{1}{6}$

23. How many different three-digit area codes can be made if repetition is allowed and the first digit cannot be 0 or 1?

(A) 800 (B) 100 (C) 80
(D) 28 (E) 1,000

24. What is the sale price of a snow blower originally priced at $370.00 and now on sale for 40% off?

(A) $330.00 (B) $222.00 (C) $148.00
(D) $14,800 (E) $22.20

25. Water is draining out of a pool at a rate of 50 gallons per minute. If there are 12,650 gallons of water in the pool, how many hours will it take to empty the entire pool?

(A) 25 hours 18 minutes (B) 6 hours
(C) 4 hours 13 minutes (D) 352 hours
(E) 253 hours

Answers and Explanations

1. A

This problem can be expressed as an expression. Since the first play started at the 22nd yard line, the expression becomes 22 + –9 (loss of 9 yards) + –5 (loss of 5 yards) + 12 (gain of 12 yards) + –2 (loss of 2 yards). Use the commutative property from chapter 1 to change the order of the numbers to 22 + 12 + –9 + –5 + –2. By combining the positive values and the negative values, the expressions yields a result of 34 + –16 = 18. After the fourth play, they were on the 18th yard line.

2. B

To find the factors of the trinomial, write the factors in the form $(x + a)(x + b)$ where the sum of a and b is 8 and the product of a and b is 16. Since the factors of 16 that have a sum of 8 are 4 and 4, the factors are $(x + 4)(x + 4)$. This is also equivalent to $(x + 4)^2$, because it is a perfect square trinomial.

3. A

First factor, and then cancel any common factors between the numerators and denominators. $\dfrac{5x - 15}{9y^2} \times \dfrac{18y}{x - 3} =$

$\dfrac{5\cancel{(x - 3)}}{\cancel{9y^2}} \times \dfrac{\cancel{18}y}{\cancel{x - 3}}$. Multiply the remaining factors to get

$\dfrac{5 \times 2}{y} = \dfrac{10}{y}$.

4. D

Using the correct order of operations that was explained in chapter 1, first divide inside the parentheses to get $-90 + 45 = -2$. Keep in mind that the result was negative because there was one negative within the parentheses. Now multiply -2 by -3 to get 6. In this step, an even number of negatives gives a positive result.

5. D

The diagram is a right triangle situation. Since one angle is known, and only one side of the triangle is known (the height of the lighthouse is opposite to the angle of $30°$), you need to use trigonometry to solve the problem. The horizontal distance from the lighthouse to the boat is the side that is adjacent to the angle of $30°$. Use the tangent ratio to solve for x, the adjacent side.

$$\text{Tan } 30° = \frac{\text{length of the opposite side}}{\text{length of the adjacent side}} = \frac{45}{x}.$$

So $x = \dfrac{45}{\text{Tan} 30°}$, or x is approximately 77.9 feet.

6. B

The area of a rectangle is $A = b \times h$. $A = 32 \times 28 = 896$ ft^2.

7. B

This represents the distributive property, choice (B), which states that multiplication distributes over subtraction. Instead of subtracting first, you can multiply 7 x 12 and then 7 x 3 and, as a last step, find the difference of these products to get the same answer. This is not an example of the order of operations, choice (A), which is a set order for performing arithmetic operations. This is not an example of the associative property, choice (C), which states that changing the grouping of either addends or factors does not affect the resultant sum or product. Likewise, choice (D) is not the correct because the commutative property states that changing the order of addends or factors does not affect the resultant sum or product.

8. E

Use the slope formula.

$$m = \frac{\text{change in } y}{\text{change in } x} = \frac{y_1 - y_2}{x_1 - x_2} = \frac{2 - (-1)}{9 - 6} = \frac{3}{3} = 1.$$

9. E

As shown below, on a coordinate grid the two equations in the system would intersect at the point (4, 9). Therefore, this is the solution to the system of equations.

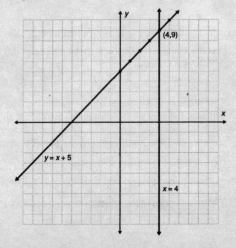

10. A

You need to follow the correct order of operations for this formula. Substitute in 104 for F, and then subtract because this expression is within parentheses. $104 - 32 = 72$. Now, mutiply this result by $\frac{5}{9}$:

$$\frac{5}{9} \times 72 = \frac{5 \times 72}{9} = \frac{360}{9} = 40 \text{ degrees Celsius.}$$

11. B

Choice (A) is not divisible by 9 because the sum of the digits, $9 + 9 + 6 = 24$, is not divisible by 9. Choice (B) fits the criteria: $1 + 7 + 1 + 4 + 5 = 18$, which is divisible by 9. Choice (C) does not work; $9 + 8 + 0 = 17$, and 17 is not divisible by 9.

12. D

First, change the mixed numbers to improper fractions: $3\frac{5}{8} = $

$\frac{(3)(8) + 5}{8} = \frac{29}{8}$; $4\frac{1}{4} = \frac{(4)(4) + 1}{4} = \frac{17}{4}$. Then change the

second fraction to have a denominator of 8: $\frac{17}{4} = \frac{34}{8}$. Then, subtract

your numerators: $\frac{29}{8} - \frac{34}{8} = \frac{-5}{8}$.

13. E

This is a decimal subtraction problem. Both numbers have two digits to the right of the decimal point, so just line up the decimal points and subtract. $19.03 - 9.99 = \$9.04$.

14. C

The sum of the measure of any two sides of a triangle must be greater than the other side. The only set that passes this condition is choice (C), $\{ 6, 8, 10 \}$. For choice (A), $6 + 7 = 13$, which is less than the third side, 14. For choice (B), $5 + 5 = 10$, which is not greater than 10.

15. A

These words translate into the two-step equation $\frac{1}{2}x + 19 = 44$.

Subtract 19 from both sides of the equation to undo addition.

$\frac{1}{2}x + 19 - 19 = 44 - 19$. This simplifies to $\frac{1}{2}x = 25$. Now, undo

multiplication by dividing both sides by $\frac{1}{2}$, which is equivalent to

multiplying by 2: $2 \times \frac{1}{2}x = 25 \times 2$, or $x = 50$.

16. D
All of the three statements are true.

17. A
First, perform multiplication in the numerator by adding the exponents.
$2^4 \times 2^5 = 2^9$. Then divide by subtracting the exponents. $\frac{2^9}{2^3} = 2^{9-3} = 2^6$.

18. E
First, simplify the square root of 150. The largest perfect square factor
of 150 is 25, so $\sqrt{150} = \sqrt{25} \times \sqrt{6} = 5\sqrt{6}$. Then, multiply this by
the coefficient of 2: $2 \times 5\sqrt{6} = 10\sqrt{6}$.

19. E
This trinomial cannot be factored, so you must use the quadratic
formula, where $a = 5$, $b = -4$, and $c = 2$:

$\frac{-(-4) \pm \sqrt{(-4)^2 - (4)(5)(2)}}{2(5)}$, which is then $\frac{4 \pm \sqrt{16 - 40}}{10}$

$= \frac{4 \pm \sqrt{-24}}{10}$. There is no real solution to the square root of a

negative number, so there is no real solution to this equation.

20. E
Since the *part* to *part* ratio of trombone players to trumpet players is 3
to 4, then the parts can be expressed as $3x$ and $4x$. Thus, the total
number of trombone and trumpet players can be expressed as $3x + 4x$
$= 7x$. The total, therefore, would need to be a multiple of 7. The only
answer choice that is not a multiple of 7 is choice (E).

21. C

Use the simple interest formula: $I = prt$, where I is the interest earned, p is the principle invested (4,500.00), r is the percentage rate (8.5%, or 0.085), and t is the time (18 months, or 1.5 years). $I = 4,500.00 \times 0.085 \times 1.5$, or $I = \$573.75$.

22. B

The prime numbers on a die are 2, 3, and 5. The probability of rolling a prime number is therefore $\frac{3}{6} = \frac{1}{2}$.

23. A

Since the first digit cannot be a 0 or a 1, there are 8 choices for the first digit. Since there are no restrictions on the second or third digit, and repetition is allowed, there are 10 choices for each of these digits. The total number of area codes is therefore: $8 \times 10 \times 10 = 800$.

24. B

Because the snow blower is 40% off of the original price, the sale price will be $100\% - 40\% = 60\%$ of the original price. Multiply: $0.60 \times 370.00 = \$222.00$.

25. C

Total Amount = rate x time. Since the water is draining at a rate of 50 gallons per minute, 12,650 gallons = 50 x time. Divide the total gallons by this rate. $\frac{12,650 \text{ gallons}}{50 \text{ gallons per minute}} = 253$ minutes.

Divide the number of minutes by 60 (to find the hour amount) and you get 4, remainder 13. This is equivalent to 4 hours 13 minutes.

SOURCE SERIES CONTEST RULES

NO PURCHASE NECESSARY.
Void in Quebec, in Puerto Rico, and wherever prohibited or restricted by law.

Kaplan Publishing wants to know how Math Source made a difference in your life. What goal(s) did you accomplish with the help of Math Source?

1) ENTRY REQUIREMENTS:
Register to enter the contest by emailing your answer to sourcecontest@simonandschuster.com or mailing your response to Source Contest, Kaplan Publishing/Simon & Schuster, 1230 Avenue of the Americas, NY NY 10020. Enter by submitting your answer as specified below.

2) CONTEST ELIGIBILITY:
This contest is open to legal residents of the United States and Canada (excluding Quebec). Employees (or relatives of employees living in the same household) of Simon & Schuster, VIACOM, or any of their affiliates are not eligible. This contest is void in Quebec, Puerto Rico and wherever prohibited or restricted by law.

3) FORMAT:
Entries must not be more than 250 words long. The author's name, address, e-mail address, and phone number must appear on the first page of the entry. All entries must be original and the sole work of the Entrant and the sole property of the Entrant.

All submissions must be in English. Entries are void if they are in whole or in part illegible, incomplete, lost, late, illegible, technically corrupted, or damaged or if they do not conform to any of the requirements specified herein. Sponsor reserves the right, in its absolute and sole discretion, to reject any entries for any reason, including but not limited to based on sexual content, vulgarity and/or promotion of violence.

4) ADDRESS:
Entries must be submitted via email by 2/1/06 to: sourcecontest@simonandschuster.com or via mail to Source Contest, Kaplan Publishing/Simon & Schuster, 1230 6th Avenue, NY NY 10020.

Each entry may be submitted only once. Please retain a copy of your submission.

5) PRIZES:
Five Grand Prize winners will receive:

Their choice of $100 worth of DVDs from Paramount Pictures. The sponsor will supply the DVD list for the winners to choose from. Total retail value of all prizes to be awarded: $500. All taxes on prizes will be the sole responsibility of the winners.

6) JUDGING:
Submissions will be judged on the originality of the story. Judging will take place on or about 3/1/06. The judges will include 5 employees of Sponsor, who are qualified to apply the aforementioned judging criterion. The decisions of the judges shall be final and binding. All prizes will be awarded provided a sufficient number of entries are received that meet the minimum criteria established by the judges.

7) NOTIFICATION:
The winners will be notified by email or phone on or about 3/15/06. All winners will be required to sign and return an Affidavit of Eligibility/Release within 30 days of notification, or prize will be forfeited. In the event any winner is considered a minor in his/her jurisdiction of residence, such winner's parent/legal guardian will be required to sign and return all required documents. In the event any prize notice is returned as undeliverable, such winner will forfeit the prize.

8) INTERNET:

If for any reason this Contest is not capable of running as planned due to an infection by a computer virus, bugs, tampering, unauthorized intervention, fraud, technical failures, or any other causes beyond the control of the Sponsor which corrupt or affect the administration, security, fairness, integrity, or proper conduct of this Contest, the Sponsor reserves the right in its sole discretion, to disqualify any individual who tampers with the entry process, and to cancel, terminate, modify or suspend the Contest. The Sponsor assumes no responsibility for any error, omission, interruption, deletion, defect, delay in operation or transmission, communications line failure, theft or destruction or unauthorized access to, or alteration of, entries. The Sponsor is not responsible for any problems or technical malfunctions of any telephone network or telephone lines, computer on-line systems, servers, or providers, computer equipment, software, failure of any e-mail or entry to be received by the Sponsor due to technical problems, human error or traffic congestion on the Internet or at any Web site, or any combination thereof, including any injury or damage to participant's or any other person's computer relating to or resulting from participating in this Contest or downloading any materials in this Contest. CAUTION: ANY ATTEMPT TO DELIBERATELY DAMAGE ANY WEB SITE OR UNDERMINE THE LEGITIMATE OPERATION OF THE CONTEST IS A VIOLATION OF CRIMINAL AND CIVIL LAWS AND SHOULD SUCH AN ATTEMPT BE MADE, THE SPONSORS RESERVE THE RIGHT TO SEEK DAMAGES OR OTHER REMEDIES FROM ANY SUCH PERSON(S) RESPONSIBLE FOR THE ATTEMPT TO THE FULLEST EXTENT PERMITTED BY LAW. In the event of a dispute as to the identity or eligibility of a winner based on an e-mail address, the winning entry will be declared made by the "Authorized Account Holder" of the e-mail address submitted at time of entry. "Authorized Account Holder" is defined as the natural person 18 years of age or older who is assigned to an e-mail address by an Internet access provider, on-line service provider, or other organization (e.g., business, education institution, etc.) that is responsible for assigning e-mail addresses for the domain associated with the submitted e-mail address. Use of automated devices are not valid for entry.

9) GENERAL INFORMATION:

All submissions, including answers, become sole property of Sponsor and will not be acknowledged or returned. By submitting an entry, all entrants grant Sponsor the absolute and unconditional right and authority to copy, edit, publish, promote, broadcast or otherwise use, in whole or in part, their entries, along with their names and addresses (city and state), in perpetuity, in any manner without further permission, notice or compensation. Entries that contain copyrighted material must include a release from the copyright holder. Prizes are non-transferable. No substitutions or cash redemptions, except by Sponsor in the event of prize unavailability.

In the event that there is an insufficient number of entries received that meet the minimum standards determined by the judges, all prizes will not be awarded. Void in Quebec, Puerto Rico and wherever prohibited or restricted by law. In the event any winner is considered a minor in his/her state of residence, such winner's parent/legal guardian will be required to sign and return all necessary paperwork.

By entering, entrants release the judges and Sponsor, and its parent company, subsidiaries, affiliates, divisions, advertising, production and promotion agencies from any and all liability for any loss, harm, damages, costs or expenses, including without limitation property damages, personal injury and/or death arising out of participation in this contest, the acceptance, possession, use or misuse of any prize, claims based on publicity rights, defamation or invasion of privacy, merchandise delivery or the violation of any intellectual property rights, including but not limited to copyright infringement and/or trademark infringement. Contest void in Quebec, Puerto Rico and where prohibited by law.

10) LIST OF WINNERS.

To obtain a list of winners, send a self-addressed, stamped envelope by 3/15/06 to: Helena Santini, Kaplan Publishing, 1230 Avenue of the Americas, NY NY 10020.

Sponsor: Kaplan Publishing, an imprint of Simon & Schuster, Inc.

1230 Avenue of the Americas, New York, NY 10020